Climate Change Education

Climate change is complex and there is a need to educate our future generations so that they are able to deal with the plethora of information and views that they come into contact with in their lives. This book inquires into what it means to teach and learn about climate change.

Now in its second edition, Chang further explores what education for climate change entails, discussing the concept of climate change education (CCE) itself, how it is taught in schools and how public education is being carried out. Featuring updated literature in a quickly advancing field, the book defines CCE for the global citizen and looks at pedagogies supporting CCE. It also identifies teachers as key stakeholders in climate change discourse, how to improve teacher readiness on the topic and how teacher professional development can support successful implementation of CCE.

This book will be invaluable to climate change educators and can act as a reference resource for teachers, education policymakers and public education agencies.

Chang Chew Hung is Dean of Academic and Strategic Development at the National Institute of Education, Nanyang Technological University, Singapore, looking after international partnerships, strategic planning, academic quality and innovations in teaching and learning for the institute. He is the Co-Editor of the journal *International Research in Geographical and Environmental Education* (*IRGEE*) and is the International Editor of the journal *Review in Geographic Education Online* (*RIGEO*) for the Asia and Oceania region. His work in sustainability education has extended beyond his research in Climate Change Education to leading the Network of Education for Sustainability in Asia, that is organised from the NTU Sustainable Earth Office (SEO). He is also an elected Senator to the Academic Council of the University and is concurrently an Adjunct Professor of East China Normal University, Shanghai, China.

Routledge Research in Education

This series aims to present the latest research from right across the field of education. It is not confined to any particular area or school of thought and seeks to provide coverage of a broad range of topics, theories and issues from around the world.

Recent titles in the series include:

Lessons from the Transition to Pandemic Education in the US
Analyses of Parent, Student, and Educator Experiences
Edited by Marni E. Fisher, Kimiya Sohrab Maghzi, Charlotte Achieng-Evensen, Meredith A. Dorner, Holly Pearson, and Mina Chun

International Perspectives on Drama and Citizenship Education
Acting Globally
Edited by Nicholas McGuinn, Norio Ikeno, Ian Davies, and Edda Sant

Meeting the Challenges of Existential Threats through Educational Innovation
A Proposal for an Expanded Curriculum
Edited by Herner Saeverot

Lived Democracy in Education
Young Citizens' Democratic Lives in Kindergarten, School and Higher Education
Edited by Rune Herheim, Tobias Werler, and Kjellrun Hiis Hauge

Equity, Teaching Practice and the Curriculum
Exploring Differences in Access to Knowledge
Edited by Ninni Wahlström

Narratives of Qualitative PhD Research
Identities, Languages and Cultures in Transition
Edited by Laura Gurney, Yi Wang, and Roger Barnard

Pragmatist Philosophy for Critical Knowledge, Learning and Consciousness
A New Epistemological Framework for Education
Neil Hooley

Climate Change Education
Knowing, Doing and Being
Chang Chew Hung

For a complete list of titles in this series, please visit www.routledge.com/Routledge-Research-in-Education/book-series/SE0393

Climate Change Education

Knowing, Doing and Being

Second Edition

Chang Chew Hung

Routledge
Taylor & Francis Group
LONDON AND NEW YORK

Designed cover image: © Getty Images

Second edition published 2023
by Routledge
4 Park Square, Milton Park, Abingdon, Oxon OX14 4RN

and by Routledge
605 Third Avenue, New York, NY 10158

Routledge is an imprint of the Taylor & Francis Group, an informa business

First edition published by Routledge 2014

British Library Cataloguing-in-Publication Data
A catalogue record for this book is available from the British Library

ISBN: 978-0-367-55503-0 (hbk)
ISBN: 978-0-367-55504-7 (pbk)
ISBN: 978-1-003-09380-0 (ebk)

DOI: 10.4324/9781003093800

Typeset in Galliard
by SPi Technologies India Pvt Ltd (Straive)

Contents

Figures

Tables

Preface

This second edition was written during exceptional times – the COVID-19 pandemic. As the world copes with the issues of the pandemic and the economic fallout that followed, the discourses on economic and social recovery could not be divorced from concepts of sustainability.

Reflecting on when the first edition of this book was written, it was also during a momentous period in Singapore's history. On the summer solstice day, 21 June 2013, Singapore recorded its historic highest 3-hour average Pollutant Standard Index (PSI) reading of 401 at noon (National Environment Agency, 2013). This was largely due to an early onset of the southwest monsoon winds in the region, bringing with it the thick smoke haze that shrouded the Riau province in Sumatra, Indonesia, due to land clearance by slash-and-burn activities. As luck would have it, a rare hailstone event hit Singapore on 25 June 2013, barely a week from the time Singaporeans were pleading for their Indonesian neighbours to douse the forest fires (Woo & Lee, 2013).

A good number of my ex-students who were teachers started messaging me through Facebook™ and asked me to explain why the two phenomena were occurring, and some even asked me if this was due to climate change. It was frustrating having to respond to more than a couple dozen of messages while trying very hard to finish the manuscript for the first edition of the book. However, I reminded myself why I wrote this book in the first place and this helped me reply to those messages patiently. Indeed, the impetus for writing this book was to advance the understanding of atmospheric science that could help elucidate what it means to teach and learn about climate change. Fundamental to this intent was to advance a clear understanding of the subject matter knowledge so that students and teachers alike can engage in climate change discourse meaningfully. Further, the curriculum development and assessment perspectives will also provide educators a framework to teach climate change education topics, which are often integrated and even disguised within existing subjects in formal curricula. In addition, public education is also discussed in the book to offer possible ways with which climate change discourse can be developed for the person on the street.

While the book has been written largely from material about climate change education in Singapore, the intention is for the local case to be compared to studies around the world, with a view to inform the teaching and learning of climate change in a wider context. As the discourse of climate change unravels with developments in the international scientific community and in response to the economic and social impact due to the reopening of borders following the easing of the global pandemic, what we should equip our learners with should not be just new facts or insights, but rather the knowledge, skills and attitudes to engage the information to make educated decisions and, subsequently, act on our common environmental future.

1 Introduction

"The science is clear: climate change is real" (Busch & Judick, 2021). What do we mean when we say we have learnt about climate change? Perhaps one of the most relevant and contemporary socio-scientific issues of our times, climate change has been shrouded in controversy, at least in the way the issue has been portrayed by mass media. In fact, the argument about climate change has evolved over the last couple of decades. From a position of denying that climate change is happening, to one that concedes that while it is happening, it may not be caused by humans, to a position that even if we are the cause of the changes, the magnitude of change may not be as dire as what has been reported. The worst is probably the latest trend where naysayers now agree that it is almost certain that climate change is caused by human activity, but pose the question of "so what" (Busch & Judick, 2021) and citing benefits such as increased food production in some parts of the world as an argument for inaction. Despite its importance in political, economic and social discourses, there is yet sufficient public interest that has been translated into pervasive, collective and sustainable action on mitigating the impact of climate change.

According to Miller (2012), people's reluctance to alter current lifestyles has been a key contributor to this apparent disengagement. Indeed, the issue is relegated as secondary when compared to more immediate and personal concerns, which unsurprisingly include issues of employment, inflation and more recently recovery from the global pandemic to a new normal. Sadly, climate change is seen as a distant threat (Chang & Pascua, 2014). O'Connor, Bord, Yarnal and Wiefek (2002) suggest that years of exposure to campaigns that invoke anxiety and fear have made the public insensitive or even resilient to alarmist discourse. The likelihood of individuals making a lifestyle change is slim unless they believe that the failure to address climate change affects their immediate worries about job security, personal freedom and the health of the economy. This phenomenon is perhaps grounded in the public's lack of accurate understanding of the issue, exacerbated by the proliferation of an apparent debate that confuses peoples' understanding of climate change.

The purpose of writing this book is to further the understanding of what it means to teach and learn about climate change. Indeed, by clarifying the subject matter knowledge, students and teachers can engage in climate

DOI: 10.4324/9781003093800-1

change discourse meaningfully. Educators will also be provided with a framework and introduced to pedagogies to teach climate change education (CCE) topics, which are often embedded within existing subjects in formal school curricula. Further, policies can be informed in developing effective public education that offers the citizenry possible ways to engage in climate change discourse. Indeed, this book, while written primarily for climate change educators, can be a reference resource for teachers, education policy-makers and public education agencies.

This chapter will open the discussion on the issue of climate change with particular emphasis on the evolution of the debate, how pervasive it has become as a global concern and the related responses of world bodies and governments to the issue. Different angles to the argument will be presented over the dispute of ideas between alarmists, contrarians and a socially circumscribed but ambivalent public. Critical points in the historical development of the discourse will be highlighted to illustrate how climate change's status has been catapulted into a legitimate crisis based on evidence of anthropogenic influence on the increasing temperatures on Earth. In order to provide authentic examples for discussion, the content of this book is largely collected from a series of research studies on CCE in Singapore. The purpose of which is to inform and contextualise and by drawing on examples from the literature on similar studies around the world, compare and suggest the way forward for teaching and learning about climate change. An analysis of the literature is also undertaken as a way to provide further comparison of the studies conducted in Singapore with those around the world.

Since the bulk of the examples used in discussions about CCE in this book are drawn from the Singapore context, this chapter will also offer a glimpse into the current state of climate change discourse and policies in Singapore. Certainly, how this informs CCE will also be discussed. We need to help learners critically engage new information about climate change as well as be empathic individuals who are committed to taking action to make their living environment a better one (Chang & Pascua, 2017).

To begin with, some people assert that climate change may not be happening. Others believe that while climate change is occurring, it may not be entirely a result of human activity. Official sources such as the Intergovernmental Panel on Climate Change (IPCC) reports become fodder for climate alarmists in supporting their claims that erratic weather patterns are a consequence of anthropogenic climate change.

Is there a climate change debate?

> There is no question that climate change is happening; the only arguable point is what part humans are playing in it.
>
> David Attenborough (2012)

The enhanced greenhouse effect is "arguably one of the greatest environmental challenges facing humankind in the 21st century" (Schreiner, Henriksen, &

Hansen, 2005, p. 3). The terms alarmists and sceptics come to mind when this statement is raised. Interestingly, there are several key discourses of climate change, which can be broadly categorised into the scientific, economic, political and ethical discourses of climate change. The complexity of the science of climate change, however, proves to be deterrence for an informed understanding of its causes, impacts and what can be done to manage it. A large text-mining study of 16,000 documents published by climate change think tanks between 2008 and 2013 showed there is an increase in discussion about the uncertainty of climate science (Boussalis & Coan, 2016). While concerns are mounting about its long-term repercussions, short-term effects are not easily observable, creating impediments to awareness-retention and diminishing the sense of urgency. Ungar's (1992) earlier analysis of global warming suggests that much like many environmental claims, the issue has to be associated with dramatic real-world events in order to gain notice. It also suffers from losing its "celebrity status" once media coverage ebbs. Over time, weak manifestations of and long lags between observable effects (Bord, O'Connor, & Fisher, 2000) give way to ambivalence as the issue is replaced by other environmental concerns. Most notably, the climate change issue became altogether redirected from a scientific discourse to an international political and economic dispute (Ungar, 1992).

Uncertainty, complexity, vulnerability and, ultimately, the authority and validity of science-based policymaking are the most frequently mentioned themes in the scientific discourse on climate change. The climate change issue is discoursed in economics through the concepts of costs and benefits of mitigation versus adaptation, and interventionist approaches are compared to free market approaches. Issues of cooperation and conflict between states frequently plague the political discourse. Finally, there is the ethical discourse of climate change where the concept of rights and responsibilities, and accountability are key in discussing climate change.

Historically, climate change discourse is not an exclusive 21st-century occurrence. The Little Ice Age (LIA) describes a period between the 16th and 19th centuries, when average temperatures in Europe were at least 1–2°C cooler than the preceding period (Matthes, 1939). The greenhouse effect as a concept was introduced in the 19th century in which the Earth was described as a "hothouse" (Fourier, 1827), probably inspired by the rising temperatures at the end of the LIA. With the end of the LIA, the rising temperature was welcomed with positive prospects, with some scholars writing about doubling carbon dioxide concentrations and its relationship to rising temperatures (Arrhenius, 1896). The discourse of how greenhouse gases can contribute to a warming Earth was deemed desirable until about the middle of the 20th century when researchers found that atmospheric carbon dioxide was not easily absorbed by oceans (Revelle & Suess, 1957), fuelling fears that the Earth was indeed warming beyond control.

By the early 1980s, the United Nations Environment Program (UNEP) and the World Meteorological Organization (WMO) established the IPCC to improve the understanding of climate science, study the impacts, adaptation and vulnerability from climate change, as well as to consider mitigation options. Since then the IPCC has produced six editions of assessment reports in these three areas, roughly once every five years, with the latest publication of the Assessment Report Six (AR6) from 2021. At the time of manuscript submission, only WG1 (The Physical Science Basis) was available. Hence, the bulk of the discussion in this book will still be based on the Assessment Report 5 (AR5) published in 2014, with reference to the earlier reports such as Assessment Report 4 (AR4) as well.

The key findings from AR5 are focused on the following central ideas. That global warming is happening and "[h]uman influence on the climate system is clear, and recent anthropogenic emissions of greenhouse gases are the highest in history. Recent climate changes have had widespread impacts on human and natural systems" (Intergovernmental Panel on Climate Change, 2014, p. 2). Many of the observed changes in the climate system since the 1950s "are unprecedented over decades to millennia. The atmosphere and ocean have warmed, the amounts of snow and ice have diminished, and sea level has risen" (IPCC, 2014, p. 2) and that these changes in the recent decades have "impacts on natural and human systems on all continents and across the oceans" (IPCC, 2014, p. 6). Similar to AR4, "anthropogenic warming could lead to some impacts that are abrupt or irreversible, depending upon the rate and magnitude of the climate change" (IPCC, 2007, p. 53). In turn, humans can respond to climate change by "adapting to its impacts and by reducing GHG emissions (mitigation), thereby reducing the rate and magnitude of change" (IPCC, 2007, p. 56). While adaptation reduces the vulnerabilities of society to climate change, together with mitigation strategies, adaptation and mitigation can complement each other and together can significantly reduce the risks of climate change (IPCC, 2014, p. 17).

Despite the more affirmative statements made in comparison to the Third Assessment Report (TAR), concluding chapters of AR4 and AR5 included key uncertainties and complexities in the causes and effects of climate change, drivers of future climate change and their effects, as well as the responses to these changes (IPCC, 2007, p. 73). Indeed, uncertainty and complexity are two of the most important ideas that represent the scientific debate on climate change most accurately and comprehensively. While we know a great deal about how the climate system works, such as how rain is formed when moist air is forced to rise up the side of a mountain, we do not know how probable this will occur for any given time. In fact, we are unable to determine the probability distribution of rainfall on the slope of a mountain or assign probabilistic values for the purposes of modelling climate. We are thus unable to reduce risk because we do not know the probabilistic values. Indeed, the uncertainty in our understanding of the climate system is a function of the complexity of the climate system.

Given this "epistemological constraint", it is unreasonable to expect that a consensus on the probable changes to weather phenomena and the associated impact on humans can be reached easily. Consequently, two epistemological views of normal and post-normal science of climate change have dominated the discourse (Davidson, 2003). The former is entrenched in the traditional objectivist and positivistic perspectives while the latter adopts a relativist position that recognises multiple valid perspectives for a given issue rather than a "single objective narrative". However, both normal and post-normal climate science are not at competition but should be used in tandem to address the issues of climate change. Indeed, when complex climate change issues require urgent resolution, multiple perspectives and observations from "farmers, fishermen, aboriginal peoples, and others who have an intimate knowledge of local climatic conditions and wildlife" (Davidson, 2003, p. 15) should be considered in combination with the predicted impact from climate models. These stakeholders represented an extended peer community, which contributes to the multiple perspectives that are a distinctive hallmark of post-normal science.

The reports produced by IPCC should be lauded for their foresight in including a large panel of authors in each of the working groups, in essence, constituting an extended peer community, which renders some degree of legitimacy to the resultant discourse in the post-normal perspective. One key concern arising out of the contrasting rhetoric between normal and post-normal climate science is to distinguish the range of impact that climate change brings to humans and society (Davidson, 2003). In particular, the economic discourse on climate change focuses on the cost and benefits arising from the effects of changing weather patterns and rising temperatures.

The early debates of climate change centred on the scientific aspects of the issue, in that the science was inconclusive about the threats to the environment and to humankind but by TAR, there is general acceptance that climate change was occurring, maybe except for the then United States Bush administration (Davidson, 2003). Once the debate on whether climate change was real started to subside, the discourse shifted to the economic ramifications of the Kyoto Protocol. Named after where it was adopted, in Kyoto, Japan, on 11 December 1997, the international agreement commits its member countries of the United Nations Framework Convention on Climate Change (UNFCCC) to targets on reducing greenhouse gas emissions. In December 2012, the Doha Amendment to the Kyoto Protocol was adopted in which countries aimed to reduce greenhouse gas emissions by at least "18 percent below 1990 levels in the eight-year period from 2013 to 2020" (United Nations Framework Convention on Climate Change, 2013). The Parties with commitments under the Kyoto Protocol (Annex B Parties) have accepted targets for limiting or reducing emissions. These targets are expressed as levels of permissible emissions, or "assigned amounts", over the 2008–2012 commitment period. The allowed emissions are divided into "assigned amount units" (AAUs). A direct consequence of this strategy to mitigate climate change has resulted in a new market for trading emissions. Emissions trading, as set out in Article 17 of the Kyoto Protocol, allows countries that have spare capacity of AAUs to sell this excess capacity to countries that have exceeded their targets (United Nations Framework Convention on Climate Change, 2013).

While the early economic discourses following the Kyoto Protocol were concerned with the loss of jobs should carbon-intensive economic activities be substituted by alternative production methods that require vastly different skill sets, the environmentalist stressed that economic impacts that will occur if the Kyoto Protocol was not implemented far outweigh the perceived loss of jobs. While the IPCC AR4 presented how major sectors of the economy can be affected by climate change impact, AR5 classified the impact on society and economy under the heading of "Water, food and urban systems, human health, security and livelihoods" (Field et al., 2021, p. 1). Table 1.1 provides a summary of these impacts. Moreover, the AR4 Working Group 2 (WGII) findings provided finer resolution examples that are shown in Table 1.1.

The findings from Working Group 2 (WGII) were instrumental in informing the adaptation and mitigation strategies suggested in AR4 and AR5, and I argue, motivated in no small part by the economic discourse on the issue. Coincidentally, at about the same time AR4 was published, the United Kingdom produced a report on the economic impact of climate change. Water, food, health, land and abrupt changes to the environment were the key areas of impact that were presented in the Stern Report of 2007. Titled the "The Economics of Climate Change", the report highlighted the stakes on a global perspective about the economics of climate change, which is an "international collective action problem" (Stern, 2007, p. iv). Indeed, the report suggests that emissions have to be cut, albeit at a cost to the economy – "around 1% of GDP by 2050 – a level that is significant but manageable" (Stern, 2007, p. xii). Sadly, the support for Kyoto Protocol in businesses is uneven. While the European business community has been "largely supportive of the Kyoto Protocol", the same cannot be said of corporations in other parts of the developed world (Davidson, 2003, p. 20) (Table 1.2).

Further, the scientific discourse about uncertainty and risks has been translated into an economic issue, especially when the manifestation of risk in the insurance industry is based on property or economic loss and the resultant increase in premiums. Adaptation strategies are often accompanied with the possibility of risk, which arises from the safe development paradoxes. While many places have started to shore up defences in preparation for sea level rise, therefore developing adaptive capacity, this might lead to a false feeling of security, which could result in greater damage and economic loss than is necessary. A classic example is hurricane Katrina which topped the charts in economic costs, with over US$200 billion worth of damage. This was due to two paradoxes, at the federal and local government levels, respectively (Burby, 2006).

Through development that makes hazardous areas safer for human occupancy, it has exacerbated the potential for economic loss through property

Table 1.1 Major projected impact on economy and society in AR5

Description of impact	*Confidence/agreement*
Food security and food production systems	
• For the major crops (wheat, rice and maize) in tropical and temperate regions, climate change without adaptation is projected to negatively impact production for local temperature increases of 2°C or more above late-20th-century levels, although individual locations may benefit	*Medium confidence*
• All aspects of food security are potentially affected by climate change, including food access, utilisation and price stability	*High confidence*
Urban areas	
• Many global risks of climate change are concentrated in urban areas. Steps that build resilience and enable sustainable development can accelerate successful climate-change adaptation globally	*Medium confidence*
Rural areas	
• Major future rural impacts are expected in the near term and beyond through impacts on water availability and supply, food security and agricultural incomes, including shifts in production areas of food and non-food crops across the world	*High confidence*
Key economic sectors and services	
• For most economic sectors, the impacts of drivers such as changes in population, age structure, income, technology, relative prices, lifestyle, regulation and governance are projected to be large relative to the impacts of climate change	*Medium evidence, high agreement*
Human health	
• Until mid-21st century, projected climate change will impact human health mainly by exacerbating health problems that already exist	*Very high confidence*
• Throughout the 21st century, climate change is expected to lead to increases in ill-health in many regions and especially in developing countries with low income, as compared to a baseline without climate change	*High confidence*
Human security	
• Climate change over the 21st century is projected to increase the displacement of people	*Medium evidence, high agreement*
• Climate change can indirectly increase risks of violent conflicts in the form of civil war and inter-group violence by amplifying well-documented drivers of these conflicts such as poverty and economic shocks	*Medium confidence*

(*Continued*)

Table 1.1 (Continued)

Description of impact	*Confidence/agreement*
• The impacts of climate change on the critical infrastructure and territorial integrity of many states are expected to influence national security policies	*Medium evidence, medium agreement*
Livelihoods and poverty	
• Throughout the 21st century, climate-change impacts are projected to slow down economic growth, make poverty reduction more difficult, further erode food security and prolong existing and create new poverty traps, the latter particularly in urban areas and emerging hotspots of hunger	*Medium confidence*

Summarised from (Field et al., 2014).

damage as it gives the government a false sense of security that steps were taken to manage the probability of damage by offering federal financial support for flood and hurricane protection works, beach nourishment and federal legislation through the "National Flood Insurance Program for safe building practices such as elevation of construction in flood hazard areas" (Burby, 2006, p. 173).

> Supposedly safe development in New Orleans (and elsewhere) has proven to be unsafe for several reasons including limitations of flood and hurricane protection works and limitations of the National Flood Insurance Program's efforts to control losses through floodplain mapping and regulation of construction practices.
>
> (Burby, 2006, p. 176)

Regardless of whether the hurricane was a direct consequence of climate change, such a simplistic consideration of the adaptation efforts to make hazardous areas safe contributes to the safe development paradox that resulted in the high economic cost and life loss in the case of hurricane Katrina, for example. This paradox is filtered to the local level where it results in a kind of "moral hazard" in which "the availability of insurance protection lowers an insured party's incentive to avoid risk" (Burby, 2006, p. 179), which does not incentivise the local governments and individuals from taking actions to manage the risk of loss and thereby reducing vulnerability.

In these economic discourses on climate change discussed above, there is a distinction between the short-term economic risk from mitigation measures versus long-term economic and ecological risk from climate change (Davidson, 2003). Clearly, international reports like AR5 and the Stern Report have set their sights on longer-term sustainability issues. From an environmental economics perspective, the Garret Harding concept of "tragedy of commons" is often used as a basis for reducing the vulnerability of

Table 1.2 Major projected impact on economy and society in AR4

Phenomenon and direction of trend	*Likelihood of future trends based on projections for the 21st century*	*Examples of major projected impacts by sector*			
		Agriculture, forestry and ecosystems	*Water resources*	*Human health*	*Industry, settlement and society*
Over most land areas, warmer and fewer cold days and nights, warmer and more frequent hot days and nights	Virtually certain	Increased yields in colder environments; decreased yields in warmer environments; increased insect outbreaks	Effects on water resources relying on snowmelt; effects on some water supplies	Reduced human mortality from decreased cold exposure	Reduced energy demand for heating; increased demand for cooling; declining air quality in cities; reduced disruption to transport due to snow and ice; effects on winter tourism
Warm spells/heat waves. Frequency increases over most land areas	Very likely	Reduced yields in warmer regions due to heat stress; increased danger of wildfire	Increased water demand; water quality problems, for example, algal blooms	Increased risk of heat-related mortality, especially for the elderly, chronically sick, very young and socially isolated	Reduction in quality of life for people in warm areas without appropriate housing; impacts on the elderly, very young and poor
Heavy precipitation events. Frequency increases over most areas	Very likely	Damage to crops; soil erosion, inability to cultivate land due to waterlogging of soils	Adverse effects on quality of surface and groundwater; contamination of water supply; water scarcity may be relieved	Increased risk of deaths, injuries and infectious, respiratory and skin diseases	Disruption of settlements, commerce, transport and societies due to flooding; pressures on urban and rural infrastructures; loss of property

(*Continued*)

Table 1.2 (Continued) Major projected impact on economy and society in AR4

Phenomenon and direction of trend	*Likelihood of future trends based on projections for the 21st century*	*Examples of major projected impacts by sector*			
		Agriculture, forestry and ecosystems	*Water resources*	*Human health*	*Industry, settlement and society*
Area affected by drought increases	Likely	Land degradation; lower yields/crop damage and failure; increased livestock deaths; increased risk of wildfire	More widespread water stress	Increased risk of food and water shortage; increased risk of malnutrition; increased risk of water- and food-borne diseases	Water shortages for settlements, industry and societies; reduced hydro-power generation potentials; potential for population migration
Intense tropical cyclone activity increases	Likely	Damage to crops; windthrow (uprooting) of trees; damage to coral reefs	Power outages causing disruption of public water supply	Increased risk of deaths, injuries, water- and food-borne diseases; post-traumatic stress disorders	Disruption by flood and high winds; withdrawal of risk coverage in vulnerable areas by private insurers, potential for population migrations, loss of property
Increased incidence of extreme high sea level (excludes tsunamis)	Likely	Salinisation of irrigation water, estuaries and freshwater systems	Decreased freshwater availability due to saltwater intrusion	Increased risk of deaths and injuries by drowning in floods; migration-related health effects	Costs of coastal protection versus costs of land-use relocation; potential for movement of populations and infrastructure; also see tropical cyclones above

Source: (Parry, Canziani, Palutikof, van der Linden, & Hanson, 2007, p. 18).

society to the impact of climate change. To some extent, the environmental effects of climate change can be seen as a market failure as the cost of production and consumption are externalised (Davidson, 2003). Environmentalists have had for a long time called for such externalities to be internalised, to put a price tag on carbon emissions, as it were.

However, the economic discourses of climate change are premised upon the scientific discourse of climate change. With the proliferation of post-normal climate science perspectives, it becomes tentative, at best, when ascribing a causal link between the enhanced greenhouse effect and the observed effect of climate change. Indeed, AR4 only went as far as to cite "almost certain" in the role of human activity in causing an enhanced warming, with the careful use of the adverb "almost", undergirding the discourse with "uncertainty". Further, the economic discourse on climate change suffers from the problem of estimating cost in the multiple emission scenarios as adopted in AR4. To complicate matters, quite a number of risks in AR5 as described in Table 1.1 were cited with medium confidence. These uncertainties fuel further debate in the political discourses on climate change.

Climate change is a "Gordian knot" of environmental problems. While international bodies like the UNFCCC struggle to disseminate the dominant discourse on climate change that is built through epistemic communities of experts, the concept of uncertainty in science and in economic costs provides fodder for contrarians to advance the debate. Scientific debates often lead to an increase in demand for more scientific research, but when used in political contexts, this uncertainty often results in a "wait-and-see" approach (Davidson, 2003, p. 30). In other words, while the concept of uncertainty affects the scientific discourse on the causes of climate change and the economic discourse on the costs of climate change, the political discourse results in varying degrees of adoption of adaptation and mitigation strategies.

While there exists a real concern about the global impact and uncertainty on the scale of effects of climate change (UNESCO & UNEP, 2011) the phenomenon has not gained enough traction to mobilise the public for a meaningful and lasting response. Differing perspectives at the eco-political level only aggravated the situation essentially denying the ordinary person meaningful comprehension of the issue (Gowda, Fox, & Magelky, 1997). As states struggle to implement strategies that adapt to the changing environments, while negotiating political commitments at an international level, the individual is often left with a barrage of information from multiple sources on the climate change topic from traditional as well as new media.

At a personal level, these debates and the knowledge on the issue must be made sense of and it would possibly inform individuals in their actions. In a broad sense, education and public education in general seem to be an avenue through which individuals engage the climate change issue. However, the position that is implicit in the rhetoric presented thus far is that climate change is happening and that climate change is unfavourable for human beings. It is therefore imperative, to determine the position of human-induced climate change within the context of CCE at the outset so that the discussion that follows can be contextualised.

I argue that CCE should be built on the foundations of moral responsibility to the environment. As much of the discussion on CCE has focused on causality as an overarching concept, it assumes that since humans are responsible for altering the climate system, all humans have to mitigate the consequential climate change impact. As there is much debate going on in the various discourses about whether the changes to the climate system observed are due, in part or entirely, to human activity, it confuses the learners and potentially erodes that obligation to act, since they may not feel guilty for the impact caused by other people after all. In considering moral responsibility, perhaps a flashback to our childhood would provide an illustration of this concept. As children, we have been taught not to waste food. The narrative presented by parents has in most cases been illustrated by images of hunger-stricken people living in a far-off country on the African continent. While some children might have asked for information and data to validate that the hunger-stricken children in Sudan suffer because of their own food wastage, most people generally accept that it was our moral responsibility not to waste food.

But moral responsibility itself is not sufficient to help our children make informed choices about what they should or should not do. For example, children may be unable to understand why they need to keep the temperature of their air conditioner thermostat at 25°C even if they know that they have a moral obligation not to waste electricity. In many cases, people tend to recall that they need to keep the thermostat at 25°C out of moral obligation rather than base that action on a deep understanding of how that action will produce results that could attenuate the impact of global warming. Unless and until the child is made to understand that power generation in their country is reliant on fossil fuel combustion, which in turn releases more greenhouse gases into the atmosphere, the reaction to being taught how to save electricity might be to simply remember the instructions. It is on this premise of developing critical thinking skills in our children that CCE is discussed in this book. CCE provides the necessary and sufficient conditions for children to act on their moral obligations.

The standard approach to teaching climate change is normally to consider *causality* as the organising concept for **why** we need to study climate change while *moral obligation* is used as a meta concept to organise **how** we teach people about climate change. However, I propose that we should begin with the concept of morality in answering the **why** or the purpose of CCE and consider causality as the organising concept for **how** to change our actions and behaviours to mitigate the impact of climate change, instead.

While the book will explore the general principles of CCE, its implementation and its effect, frequently referring to examples from around the world, the examples are grounded in the findings from empirical studies conducted in Singapore. There is a need to contextualise the responses to climate change in the Singapore example, including how CCE is conducted. With this as a backdrop, the discussion will then extend to the discussion on CCE in general.

Climate change and Singapore

Singapore is an island city-state of 728.6 km^2 with a population of 5,453.6 as of June 2021 (Department of Statistics Singapore, 2021). Although it has very little natural resources and limited land, through its good governance, well-planned economic growth and holistic urban development (Armstrong & Read, 1998; Shahid & Nabeshima, 2012) Singapore has become an affluent society with a per capita GDP of S$82,503 (US$60,222) (Department of Statistics Singapore, 2021). Singapore is very much aware of its limited resources, the high rate of economic development (Chua, 2007) and high population in an island state of very small size (Henderson, 2012).

As a nation, Singapore has responded well in creating governmental and social structures that encourage mitigation practices at national, local and individual levels (Chang, 2008). This is in alignment with a recent trend, at least for the last decade in the way the country views climate change. What seemed to have caused this current trend would be, I argue, due in no small part to the increasing occurrence of erratic weather patterns that have become featured more frequently in mass media.

Singapore is located just one and a half degree north of the equator. Textbooks on meteorology would assert that tropical cyclones or typhoons as they are known in the region, are highly unlikely (almost impossible) storms near the equator. However, a unique storm, Typhoon Vamei, made landfall in Singapore on 27 December 2001. Within a 24-hour period, the rainfall received was 240 mm or 10% of that of the total rainfall received in the entire year (Chang, 2010). The weak Coriolis force at low latitudes is the main reason why typhoons very rarely form near the equator. Scholars posit that the abnormal cyclone formation was due to the abnormally long-lived presence of a north–south pressure gradient across the equator created by a storm surge in the South China Sea.

For the tiny island state in the doldrums, this was an exciting meteorological event. In the next few years, extreme weather events that were often accompanied by heavy rain became noticed more by the public as mass media became more diligent in reporting these events. In a study of the local meteorological data over a 30-year period and a meta-analysis of studies in the literature on storm events in the Southeast Asian region, the predicted increase in frequency of high precipitation events by the IPCC AR4 was verified for Singapore (Chang, 2011). Indeed, several high precipitation events have punctuated the second half of the first decade of the 21st century. An example of such a storm was the event in December 2006. Chatterjea (2009) reports that 366 mm of rain fell over a period of 24 hours on 19 December 2006. This exceeded Singapore's 30-year average (1978–2007) monthly rainfall of 299 mm for December by close to 50%.

It is therefore unsurprising that government leaders in Singapore suggest that climate change "has serious security implications" (Bahrawi, 2007). At the Singapore Energy Conference on 4 November 2008 held in conjunction with the inaugural International Energy Week, the late Minister Mentor Lee Kuan Yew emphasised the need for Singapore to stay green and clean. This is important not just to maintain its status as a clean and green city, but Singapore "will lose business and lose our extra premium for being an unusual city" (SG Press Centre, 2008). Further, the then Prime Minister, in September 2008 in a dialogue, reflected on the difficulty of tackling the problem of climate change, which is a tricky one for leaders. He said,

> It is a long term issue, yet societies were not set up to deal with problems "that have 100-year time spans." In addition the solutions are neither cheap nor painless: the remedies to this problem also take 30 or 100 years to get moving.
>
> (Fang, 2008)

Singapore ranks 126th of 142 countries in terms of CO_2 emissions per dollar GDP, based on International Energy Agency (IEA) data (NCCS, 2022). This rough index masks the fact that Singapore is a manufacturing base for multinational corporations. Minister Mentor Lee explains that this large index was not due to the population using it, but rather the world population because the world market is using Singapore through these multinationals to produce the emissions. With the prospect of moving from big bulky industries to smaller ones such as electronics, pharmaceuticals and more into services, he opined that the ratio would go down (Koo, 2007). Deputy Prime Minister Jayakumar (at the time) explains that Singapore has laid out guiding principles with regard to the post-2012 climate change deal. One of these is to support plans to fight smoke haze and reduce emissions from the razing of forests (The Straits Times, 2008).

Singapore's National Climate Change Strategy report produced in February 2008 described strategies for climate change vulnerability and adaptation, as well as mitigation of greenhouse gas emissions (Ministry of Environment and Water Resources, 2008). In a response to a letter in the press, Director (Strategic Policy) Ministry of the then Environment and Water Resources (MEWR) stated that Singapore has set a national target of reducing the carbon intensity (carbon dioxide emissions per dollar of GDP) by 25% from 1990 to 2012. This ensures that as Singapore develops, the carbon intensity is lessened, thus balancing the economic and environmental goals of sustainable development. It has put in place a range of policies and measures to reduce carbon dioxide emissions. Promoting energy efficiency and encouraging the use of less carbon-intensive fuels in different sectors of the economy, such as industry, buildings and transport are strategic priorities for Singapore. To this end, the government has committed $350 million towards research and development, test-bedding and pilot projects in clean energy (National Research Foundation, 2008).

A major study has also been commissioned by the government to look into how global warming and climate change will impact Singapore. The two-year study by the National Univeristy of Singapore's Tropical Marine Science Institute aims to better understand what the possible long-term effects could be (The Straits Times, 2007). Even so, the impact on the local environment has become more noticeable in recent years. Flooding had affected several areas of Singapore but the construction of drainage facilities over the last three decades has largely solved these recurring problems. However, with an anticipated rise in sea level by 59 cm the vulnerability to flooding will increase. In anticipation of this, since 1991 the Public Utilities Board (PUB) has made it mandatory that "new reclamation projects to be built to a level 125 cm above the highest recorded tide level" (Ministry of Environment and Water Resources, 2008).

Sea level rise has been believed to impact coastal land areas and "70% to 80% of Singapore's coastal areas have hard wall or stone embankments, which help protect against coastal erosion, the rest are … natural areas such as beaches and mangroves" (Ministry of Environment and Water Resources, 2008). Some of the recreational areas such as East Coast park, Sungei Buloh, Pasir Ris park, West Coast park and Sentosa will be especially affected by coastal erosion due to sea level rise (Ministry of Environment and Water Resources, 2008). Moreover, Wong (1992) studied the possible level of sea level rise in Singapore pointing out the impact on natural coast, reservoirs and drainage (Wong, 1992). Measures will be taken to strengthen revetments and reduce coastal erosions and land loss as and when the need arises.

Singapore is diversifying its water sources and one of the major supply sources is the construction of coastal reservoirs protected from the sea by barrages. There is even consideration for the construction of polders as a possible future sea level rise adaptation strategy (Lee, 2019). A significant increase in sea level can cause seawater to flow into such reservoirs or seawater intrusion. However, this is unlikely to happen as the "reservoir dams are much higher" than the projected rise in sea level and the gate structures were designed to be raised during such an event. One concern with climate change is the change in rainfall patterns and a decrease in rainfall may affect the capacity of water stored in the reservoirs. However, this could be mitigated by other sources of supply including NEWater and desalination (Ministry of Environment and Water Resources, 2008).

The inherent urban heat island effect of Singapore's climate and warmer temperatures brought about by global warming would significantly increase heat stress and therefore the use of air-conditioning and energy consumption. One way which has been ongoing is to increase the amount of greenery in the city and rooftops, "using building materials with better thermal properties", lighter tonal building surfaces, layouts and maximising wind and ventilation effects (Ministry of Environment and Water Resources, 2008). In fact, Singapore prides itself as a city in the garden and this deliberate policy of urban development has caught the attention of other cities that want to model their development after the Singapore eco-city concept. Singapore's aim is to be a "city of gardens and water". On the Prime Minister's visit to Guangzhou, Nanning and Chengdu, China, he noted all three cities have been beautified. "They have seen what Singapore has done and want to try to do the same" (Lau, 2006).

Increased energy demand through the use of air-conditioning in homes and buildings must be curtailed through greater energy efficiency in buildings. Already under the Building Control Act, "air-conditioned buildings must be designed with a high-performance building envelope that meets the prescribed Envelope Thermal Transfer value (ETTV)" (Ministry of Environment and Water Resources, 2008), set at 50 W/m^2. A study was being conducted to review the ETTV standards and to apply these to residential buildings. Since 2008, buildings are awarded BCA Green Mark rating based on energy efficiency, water efficiency, site/project development and management, good indoor environmental quality and environmental protection and innovation. Points are given for each of the categories. In addition to achieving the minimum points in each rating scale, all requirements have to be met with a minimum score of "50 per cent of the points in each category" (Ministry of Environment and Water Resources, 2008).

Competition for scarce resources in a global warming world and the possible displacement of peoples across borders can potentially "deepen tensions, and provoke conflicts and wars" (Bahrawi, 2007), which raises severe concerns for a small island state like Singapore. While Singapore is an affluent nation, it is small and susceptible to changes in the global market. Global-warming-induced climate change can result in economic and social impact that will consequently affect Singapore and its people. It was set against such concerns that the NCCS was set up as part of the Singapore Green Plan 2012 to respond to climate change issues.

> Singapore is aware of its vulnerability to climate change. To be better prepared for these possible climate change impacts, the Government is commissioning studies to better understand the detailed impacts of climate change on Singapore, which will facilitate the planning of adaptation measures. Government agencies will work together to establish national systems to actively monitor and manage these vulnerabilities, and where necessary, take precautionary measures.
>
> (Ministry of Environment and Water Resources, 2008)

In all, Singapore as an island state is vulnerable to climate change and the potential impact reported in the Singapore National Climate Change Strategy Report include the following:

a. Increased flooding
b. Coastal land loss
c. Water resources scarcity
d. Public health impact from resurgence of diseases
e. Heat stress

f. Increased energy demand
g. Impact on biodiversity

(Ministry of Environment and Water Resources, 2008)

In 2006, the country of 5 million residents became the 168th state to ratify the Kyoto Protocol (Ministry of Environment and Water Resources, 2006). Singapore further ratified Doha Amendment to Kyoto Protocol in 2014 and signed the Paris Agreement on 22 April 2016 and ratified it on 21 September 2016. Since then, Singapore has also taken part in various multinational initiatives that encourage comprehensive and holistic approaches to climate change mitigation. These include meetings, talks and conventions under the World Trade Organisation (WTO), World Intellectual Property Organisation (WIPO), International Maritime Organisation (IMO) and International Civil Aviation Organisation (ICAO). Environmental cooperation has also been actively pursued through bilateral and regional platforms, such as the Sino-Singapore Tianjin Eco-City with China, the Asia-Pacific Economic Cooperation (APEC), the Association of Southeast Asian Nations (ASEAN) and the Jambi Province to improve peat land management and promote sustainable land use – with Indonesia (National Climate Change Secretariat, 2022).

The government has since made known its intentions to join prevailing discourses on greenhouse gas emissions management. From a position of ambivalence, it now sees irregularities in the climate as "one of the most pressing environmental challenges of the global community" (Ministry of Environment and Water Resources, 2006, p. 26). This shift in perspective coincides with rising global concern about climate change, as well as developing local empirical data on climate change's impacts on the island. Chang's (2011) analysis, for instance, illustrates the vulnerability of Singapore in terms of observed increases in temperature and precipitation (Chang, 2011). Several studies also singularly identified sea level rise as the main threat to Singapore (Wong, 1992; Hussain, 2007; Gunasingham, 2010), impacting the island nation's waterways, coasts and drainage system.

Ng and Mendelsohn (2005) document that Singapore invests heavily in securing its developed and reclaimed lands. The SGP discusses these adaptive and pre-emptive measures in place. However, even with a seemingly formidable system in place, the vulnerability of Singapore to climate change still often becomes eventfully apparent and one which catches everyone by surprise. One such event is the inundation of Orchard Road, the main retail and business district of the country, on the 17 June 2010 (The Straits Times, 2010). The government responded to the crisis through structural engineering to prevent a similar scenario from happening, leading to a comprehensive risk mapping of areas threatened by flooding and sea level rise.

While Singapore has ratified the Kyoto Protocol, environmentalism, environmental education and action in Singapore are not new to the people. For a first-time visitor to the city-state, care for the environment in Singapore is evidently manifested in keeping an ultra-urban, cosmopolitan island as a "garden city", a greening scheme by the government introduced in 1965

(Kong et al., 2000). In response to the rapid urbanisation in the decades preceding this move, the greening scheme, which was really a managed messicol vegetation, was seen as a remedy to the loss of natural environment (Kong, Yuen, Sodhi, & Briffett, 1999). From its early days of independence to the late 1980s, green environmental action was literally concerned with planting trees and creating a garden city. It was not until the early 1990s that the first Singapore Green Plan (SGP) was unveiled at the Earth Summit in Rio de Janeiro, Brazil, in June 1992, targeting broader environmental issues such as pollution. By 2002, Singapore introduced a new green plan called SGP2012, which was subsequently revised in 2006, detailed plans to attain a vision of sustainability for 2012, including targets for reducing carbon emissions (Table 1.3).

This was Singapore's ten-year blueprint towards environmental sustainability (Ministry of Environment and Water Resources, 2006). Views from more than 17,000 individuals were sought in a review where key environmental issues like climate change were raised. The SGP2012 (2006 edition) covers six focus areas: Air and Climate Change, Water, Waste Management, Nature, Public Health and International Environmental Relations. It was not until the 2006 version that issues of climate change were brought to the foreground, with a goal "[t]o improve carbon intensity (i.e. CO_2 emission

Table 1.3 Some of the targets set in the updated SGP 2012

Focus area	*Selected targets*
Air and climate change	Maintain the Pollutant Standards Index (PSI) for ambient air within the "good" range for 85% of the year and within the "moderate" range for the remaining 15%
	Reduce the ambient Particulate Matter 2.5 (PM 2.5) level to within an annual average of 15 μg/Nm³ by 2014
Water	Increase catchment areas to 67% of Singapore's land surface
	Increase the supply of water from non-conventional sources, such as desalination and water reclamation, to at least 25% of Singapore's water demand
	Reduce per capita domestic water consumption to 155 litres a day by 2012
Waste management	Increase the overall waste-recycling rate to 60% by 2012.
	Extend the lifespan of Semakau Landfill to 50 years, strive towards "zero landfill" and "close the waste loop"
Conserving nature	Establish more parks and green linkages
	Set up a National Biodiversity Reference Centre
Public health	Increase community ownership to sustain a high standard of public health
	Maintain low incidence of vector-borne and food-borne diseases.
International environmental relations	Intensify collaboration with partners at regional and global levels to tackle environmental challenges

Source (Chew, 2008).

per GDP Dollar) by 25% from 1990 level by 2012" (Ministry of Environment and Water Resources, 2006, p. 9). To do so, three key strategies were proposed:

1. Promote energy efficiency
   - Develop enablers to make consumers more energy efficient/fuel efficient
   - Improve energy management practices of businesses
2. Promote use of clean energy
   - Promote use of cleaner energy (such as natural gas)
3. Demonstration projects on renewable energy
   - Encourage demonstration projects on renewable energy (such as solar and biomass)

(Ministry of Environment and Water Resources, 2006)

By 2006 there was greater acceptance of the TAR findings at international levels, in conjunction with the increased media coverage on extreme rainfall events and possibly the environmental costs it has brought, the SGP2012 recognises that greenhouse gases (mainly carbon dioxide) have contributed to climate change. Although SGP2012 proposes that Singapore only contributes to less than 1% of the global CO_2 emissions, the government seeks improvement in terms of the carbon intensity, or CO_2 emission per GDP Dollar (Ministry of Environment and Water Resources, 2006). It is interesting to note that the targets set for reducing carbon emissions are defined by monetary terms, perhaps in no small part due to the dollar value tagged to the environmental costs of the 2006 heavy rainfall event discussed above.

In 2015, the country introduced the Sustainable Singapore Blueprint 2015 which builds on the 2012 plan and in particular focuses on five strategies:

1. Eco-smart endearing towns
2. "Car-Lite" Singapore
3. Towards a zero waste nation
4. Leading green economy
5. Sustainable housing

(Ministry of Environment & Water Resources & Ministry of National Development, 2016)

These strategies further complement the aspirations set out in the SG2012 by setting goals such as "achieving a peak hour mode share of 75% for public transport by 2030, up from 64% in 2013" and "expanded use of solar energy". Between 2015 and the new Green Plan announced in 2020, Singapore also adopted the Katowice Climate Package which included a set of guidelines and procedures relating to the implementation of the Paris Agreement, in 2018.

In his National Day rally speech in 2019, Singapore Prime Minister Lee Hsien Loong focused his speech on the plans to mitigate and adapt to the impact of climate change. To understand the climate change issue, he announced plans to set up the Centre for Climate Research that brings together a multidisciplinary team of scientists to understand how climate change is affecting Singapore and the neighbouring countries in Southeast Asia. To mitigate climate change and as part of the Paris Climate Agreement, Singapore committed to slowing and eventually capping CO_2 emissions by 2030. A carbon tax was proposed as a means to promote sustainability and mitigate climate change. Concrete examples of how some MRT stations will be built with elevated entrances and plans for flood protection for the entire MRT system were announced to adapt to climate change. New developments will also be constructed at higher elevations, and coastal defences will be constructed to protect low-lying areas. The construction of polders and seawalls was also mentioned as a way to adapt to the effects of climate change on sea level rise (Lee, 2019).

Singapore also pledged to reduce greenhouse gas emissions by 16% below business as usual (BAU) levels in 2020 and made further commitment to reduce Carbon Emission intensity by 36% from 2005 levels by 2030. This is to stabilise greenhouse gas emissions with the aim of peaking around 2030. Singapore submitted its enhanced Nationally Determined Contribution (NDC) and Long-term-low-Emissions Development Strategy (LEDS) document to the UNFCCC on 31 March 2020. This NDC sets a target for absolute emissions to peak at 65 Mt CO_2 emission by around 2030. The LEDS builds on enhanced NDC by aspiring toward net zero emissions as soon as viable in the second half of the 21st century. Singapore also set up an Inter-Ministerial Committee on Climate Change to ensure that Singapore adopts a coordinated approach towards climate change with regard to the LEDS and NDC (National Climate Change Secretariat, 2022).

In 2020, Singapore also announced a refreshed green plan called the Singapore Green Plan 2030 (SGP2030). This is timely given that AR5 was published after the SGP 2012. The SGP2030 is the nation's sustainability movement that aspires to rally bold and collective action to tackle climate change. There are six key strategies in the refreshed SGP2030, namely:

1. City in Nature
2. Sustainable Living
3. Energy Reset
4. Green Economy
5. Resilient Future
6. Green Government

(Singapore Green Plan, 2021)

Among the initiatives that have been planned include planting 1 million more trees in Singapore, introducing an Eco Stewardship Programmes to enhance environmental education in schools, reducing waste to landfill per capita per day by 30% by 2030, quadrupling solar energy deployment to 1.5-gigawatt peak by 2025, funding dedicated to coastal and drainage flood protection measures and to provide leadership in green governance. These are indeed concrete plans, with specific outcomes in mind. For green governance, significant developments were witnessed in 2021 with the public sector leading the way to pursue sustainable development with Greengov.sg initiative (National Climate Change Secretariat, 2021). Minister of National Development, Mr. Desmond Lee specifically mentions an "80-80-80 in 2030" which signifies goals "to green 80% Gross Floor Area of buildings in Singapore by 2030, to have 80% of new buildings as super low energy buildings from 2030 and to have 80% improvement in energy efficiency compared to 2005 levels by 2030" (Lin, 2021). To support these goals, some of the enablers include the implementation of the SG Eco Grant Call with a total of 37 projects that will obtain funding for smaller community projects to create solutions for a sustainable Singapore (Ministry of Sustainability & Environment, 2022).

By and large, these efforts are largely state-led and state-driven. The city population has not been fully participatory. In 2009, a media-led survey found that half of the Singaporean population is not aware, and marginally know the specificities of climate change. Additionally, their understanding of their roles in mitigation appears to be undecided (Mehlsen, 2009). In the most recent study by the National Climate Change Secretariat (2012), it appears that Singaporeans are now more receptive to issues on climate change. Generally, the study found that the public was concerned about climate change and wanted more information about the subject, especially on the impacts it holds for Singapore, what the government is doing for mitigation and adaptation as well as what individuals could do to address climate change; 86% feel that they have a part to play in taking action on climate change. More than half feel that their actions could make a difference and many are contributing in various ways (National Climate Change Secretariat, 2011).

While the inherent need to have an educated citizenry exists as spelt out in the SGP and the Singapore National Climate Change Strategy, education in this sense is inferred more in the informal areas such as students' co-curricular activities in projects organised by the MEWR with non-government organisations (Ministry of Environment and Water Resources, 2008). The formal curriculum treats climate change as interspersed topics within subjects in the Natural and Social Sciences, Moral Education and Values. Together with general environmental issues, climate change issues are lumped together to form an all-purpose course, packaged as either Education for Sustainable Development (ESD) or Environmental Education (EE). By 2009, the government realised the need for coordination among government agencies, and a National Climate Change Secretariat was set up under the auspices of the Prime Minister's Office to this end. While the government has put in place a series of structures and outlined key strategies to help Singapore manage the climate change issue, there is a general lack of empathy from the people.

As Singapore becomes developed, prosperity has started to influence the values and attitudes of the citizenry towards their appreciation of the environment. While the population dwells at a confluence of commerce, trade and culture, they are increasingly looking at wealth as an indicator of success (Jiuan, Wirtz, Junk, & Keng, 2001). Although the citizens are cognisant of the environmental challenges Singapore face, those who step forward to help are likely to do so only if their personal convenience is not jeopardised. This transaction of responsibility is premised on causality rather than on moral responsibility. There is a general belief that businesses are responsible for environmental degradation and consequently, these corporations should bear the sole responsibility of mitigating the negative impact of climate change (Jiuan, Wirtz, Junk, & Keng, 2001).

Singapore is a small nation but economically significant and highly exposed to international trade. As such, it is highly sensitive to the challenges it faces to remain competitive. While in principle the policy of the government towards the environment is very much in line with the statements from the world bodies, its interpretation and translation in the national context are often determined by national priorities, challenges, interests and constraints.

Pragmatism is the apparent key determinant for environmental action in Singapore, both at a state and at the individual level. Indeed, the response is

> not necessarily because it wants to save the planet, but because it has realised that it is in its own self-interest to reduce carbon emissions, to eliminate wasteful use of energy in all its forms, and to harness green growth opportunities.
>
> (Figueres, 2012)

At an individual level, a person is unlikely to recycle, unless he or she sees some economic benefit in doing so.

Singapore youths have exhibited little interest in nature, due to their limited contact with the natural environment, having over-protective parents as well as exposure to a range of other recreational and entertainment options (Kong, Yuen, Sodhi, & Briffett, 1999). It is therefore surprising, and even ironic, that a study by the National Environment Agency found that students have high environmental awareness (National Environment Agency, 2006). Indeed, Singapore students consider ozone depletion and global warming as environmental problems that they are of utmost concern (Tan, Lee, & Goh, 1998).

Notwithstanding their understanding of the environmental issues, students remain unwilling to volunteer their free time for environmental action nor are they willing to buy recycled products at a higher cost. Similar to adult Singaporeans, many young people believe that it is the government's responsibility to keep the environment clean and green (National Environment Agency, 2006). There is certainly a socially and culturally embedded indifference to environmental concerns that has shaped the way the youth view climate change in Singapore. It is therefore unlikely that any environmental action can be effectively anchored on citizens' individual conscience and that the strong hands of the state may be required for the effort to be sustained (Kong et al., 2000).

Given the issues of climate change communication around the world and the context for climate discussions in a small island state like Singapore, we need to consider what the intent of CCE is and how we can leverage on CCE for our environmental future.

Climate change and what it means for education

In summary, climate change is complex and there is a need to educate future generations so that they are able to engage critically in the flood of information and perspectives that they encounter. Not only do we want an educated citizenry that acts to protect and conserve the quickly diminishing environmental wealth of our planet, but we also want our children to be able to engage in critical and thoughtful inquiry about the information that they receive. It is through education that we can ensure that there is a sustainable future for the Earth. It is therefore imperative that we examine what constitutes CCE.

The "potential of education is enormous…Education not only informs people, but it can also change them. As a means for personal enlightenment and for cultural renewal, education is not only central to sustainable development, but also humanity's best hope and most effective means in the quest to achieve sustainable development" (UNESCO, 2012). Until about the turn of the first decade in the 21st century, ESD has been one of the key educational approaches used to teach people about climate change. Sustainability education, Education for Sustainability (EfS) and ESD are interchangeable terms describing the practice of teaching sustainability.

In defining CCE, ESD and Environmental Education concepts must be discussed and the difference between these should be disambiguated. ESD, for example, is highly debated and contested and it means different things to different people depending on their orientations and concerns (Bonnett, 2002) and (Fien, Heck, & Ferreira, 1997). Even the meaning of sustainable development and sustainability remains a nebulous concept that is variously defined, debated and deconstructed (Higgit, 2006).

Nonetheless, ESD is broad and focuses on development issues, while this book is specifically about education for climate change. There is therefore a need to discuss at length what CCE entails, including advancing a framework that describes the subject matter, the instructional approaches and ways learning can be assessed. It is time to address the issue since climate change on Earth and its discourse is occurring at an unprecedented rate, scale and complexity.

2 Defining climate change education for the global citizen

Governments and non-governmental organisations (NGOs) have long regarded education as a primary strategy for raising environmental awareness and instilling global citizenship (Schreiner, Henriksen, & Hansen, 2005). In 1992, the United Nations Conference on the Environment and Development reaffirmed the importance of education in providing the skills and knowledge needed by the world's youth to manage the environment in which they live. In December 2002, the United Nations adopted Education for Sustainable Development (ESD) as the framework with the United Nations Educational, Scientific and Cultural Organization (UNESCO) as the lead agency. However, education faces specific challenges across contexts, at the global, national and local levels. Despite the prevalence of the concept of global citizenry in these discourses, the scale at which these settings determine how climate change education (CCE) may be carried out must be considered.

While there have been many public events and campaigns through mass media aimed at raising awareness of climate change, there is little empirical research published to show that such campaigns were efficacious in bringing about behavioural change. In addition, barriers such as being sceptical to and unwilling to inquire further about the information made available to the public might also lead to such a situation. Thus, it is argued that such barriers should not exist in the first place and one way to address this problem is through formal CCE in schools. The objective of CCE is not merely to educate citizens to be the future custodians of the environment but to also enable them to engage in critical and thoughtful inquiry about the information they receive. In the current climate of "fake news" and disinformation, educating students with the appropriate thinking skills is crucial to ensure the sustainable future of our Earth. In addition, there is the question of what CCE means for a global citizen. Many of the impacts of climate change are felt at the local level and this poses additional challenge for CCE as learners need to make sense of the changes they see at the local or national scale with the reports they read about climate change impact at the global level. This chapter will provide the conceptual framing against which the discourses on CCE can take place. What sort of learning outcomes will we want to see in our students and how can the curriculum be designed to meet these needs? The chapter will also discuss the concept of CCE across the scales of global,

DOI: 10.4324/9781003093800-2

national and local contexts. As the title of the book alludes to education beyond knowledge to skills to dispositions, the context within which a child learns about climate change becomes important in our discussion.

The trend of incorporating or integrating global CCE aims into national policy has clearly seen an upward trajectory during the last two decades. For example, Olsson, Gericke, and Chang-Rundgren (2016) emphasised that sustainable development (with climate change being a topic in the national curriculum) needs to be described for each of its dimensions with respect to their interrelation for space and time. Marginson and Rhoades (2002) presented the notion of global agency heuristics as a useful analytical framework for understanding emergent dynamics and how global discourse on CCE might affect education reforms in various locations. The framework provides a multi-"exploration and analysis of types and patterns of influence and activity, to conceptualize social relations and actions globally, nationally and locally" (Marginson & Rhoades, 2002, p. 290). In addition, it also considers the influence on local schools in terms of accountability, teacher training and curriculum implementation (Bengtsson & Östman, 2014; Aikens, McKenzie, & Vaughter, 2016; Abalahin & Chang, 2020).

In essence, Marginson and Dang (2016) extended Vygotsky's sociocultural theory of learning and applied it at multiple scales, including local, national and global, to include macro-historical changes and global relations, as well as changes at the level of individuals in society, to examine "their modes of thought, imagined possibilities, and agency freedoms" (Abalahin & Chang, 2020, p. 20). In other words, an argument can be made that the framework of the global agency heuristic allows us to examine the policy and programmatic curriculum of CCE by understanding the agency of various social elements (e.g., curriculum designers, teachers, students) at the intersection of global, national and local dimensions. As a result, the term "glonacal" citizen as used in the literature, acknowledges that CCE for a person must be taken into account across these scales.

Education for sustainable development and environmental education

CCE can be tackled under the broad umbrella of two prevailing paradigms: Environmental Education (EE) and ESD. There are conceptual overlaps between EE, ESD and CCE. However, the manner in which they are being grouped together into a single framework will prove problematic in the long run. At the very least, these two major frameworks that are used to guide the place of climate change in curriculum in regions across the world must be understood.

Education for sustainable development

Understanding what is meant by sustainable development is crucial to understanding ESD as a framework. By pursuing sustainable development, we may meet current demands without jeopardising future generations' capacity to meet their own needs (UNESCO, 2012). This development approach takes into account environmental problems as well as issues like poverty, equality, human rights, cultural diversity and universal education. And it is through education that sustainable development may be realised. It enables individuals to develop the knowledge, attitudes and skills necessary to take action, independently and collaboratively, locally and globally, to improve their overall well-being now without jeopardising our future world's environmental well-being.

Over the last few decades, public awareness of climate change, as well as education of students on climate change causes, effects and management, has increased (Padpadimitriou, 2004). There has been growing recognition of the threats brought about by climate change and it draws both citizens and politicians to search for ways to address the issue. Increasingly, educators have been encouraged to include and integrate topics on climate change into existing education and outreach programmes (UNESCO & UNEP, 2011). The UNESCO and UNEP (2011) maintain that CCE must be framed strategically to promote mitigation and adaptation with emphasis on the causes and impacts, as well as needed "knowledge, skills, values and attitudes for effective mitigation using appropriate action-oriented pedagogies" (UNESCO & UNEP, 2011, p. 55).

While there is a broad consensus on the place of education in advancing the discourse on climate change and hopefully bringing about changed behaviour in our youth, the topic has commonly been tackled through an interdisciplinary approach. This approach requires teachers to understand the complexity of the science of climate change and integrate these into the pedagogies of various disciplines. Within the framework, ESD:

- is based on the principles and values that underlie sustainable development;
- deals with the well-being of all four dimensions of sustainability – environment, society, culture and economy;
- uses a variety of pedagogical techniques that promote participatory learning and higher-order thinking skills;
- promotes lifelong learning;
- is locally relevant and culturally appropriate;
- is based on local needs, perceptions and conditions, but acknowledges that fulfilling local needs often has international effects and consequences;
- engages formal, non-formal and informal education;
- accommodates the evolving nature of the concept of sustainability;
- addresses content, taking into account context, global issues and local priorities;
- builds civil capacity for community-based decision-making, social tolerance, environmental stewardship, an adaptable workforce, and a good quality of life;
- is interdisciplinary. No single discipline can claim Education for Sustainable Development for itself; all disciplines can contribute to Education for Sustainable Development.

(UNESCO, 2012)

Indeed, with ESD as the framework, CCE is intended to be approached in a cross-curricular manner. UNESCO and UNEP specifically emphasised the need to diverge from the exclusivity of climate science upon which CCE is often packaged with. The proposal was to implement these changes across educational plans and policies, down to education systems encompassing programmes, practices, policies, assessment and teacher education (UNESCO & UNEP, 2011). However, the framework for ESD is really a list of characteristics rather than a conceptual framework for understanding an issue such as climate change. Rather than having broad descriptive features, a framework for CCE needs to address issues of "what to teach", "how to teach", "who to teach" and "how do you know they have learnt".

Environmental education

The Belgrade Charter on a global framework for environmental education was developed to guide principles for global environmental education since the 1970s. Arising from recommendation 96 of the Stockholm conference on the Human Environment that the "development of environmental education as one of the most critical elements of an all-out attack on the world's environmental crisis", the framework calls for global development ethic at the personal level. The idea is premised on educating the youth of the world to ensure a long-term solution to environmental problems (The Belgrade Charter, 1975). The goal of environmental education is thus to "improve all ecological relationships, including the relationship of humanity with nature and people and with each other which will ensure the improvement of individuals' quality of life and to ensure preservation and improvement of humanity's potentials".

The key feature of the framework is an articulation of the six objectives of Environmental Education, namely:

1. Awareness: to help individuals and social groups acquire an awareness of and sensitivity to the total environment and its allied problems.
2. Knowledge: to help individuals and social groups acquire basic understanding of the total environment, its associated problems and humanity's critically responsible presence and role in it.
3. Attitude: to help individuals and social groups acquire social values, strong feelings of concern for the environment and the motivation for actively participating in its protection and improvement.
4. Skills: to help individuals and social groups acquire the skills for solving environmental problems.
5. Evaluation ability: to help individuals and social groups evaluate environmental measures and education programmes in terms of ecological, political, economic, social, aesthetic and educational factors.
6. Participation: to help individuals and social groups develop a sense of responsibility and urgency regarding environmental problems to ensure appropriate action to solve those problems.

(The Belgrade Charter, 1975)

There are three interesting points of interest about these six objectives. Firstly, they are organised in some semblance of increasingly complex cognitive to affective and actionable abilities. This is not different from the dominant behavioural paradigm in education discourse in the 1970s. Following Bloom, Anderson & Krathwhol's work in 1956 on a taxonomy of educational objectives, behaviourists believed that when the conditions of learning are right (even replicated in some instances), learning outcomes as described in the educational objectives should follow. Much like a laboratory mouse perfecting a maze in record time to get to the cheese after repeated attempts, the learning theorists of the 1970s focused on prescribing educational objectives that guide the design of instruction. Secondly, the six Environmental Education objectives transcend beyond cognitive engagement to action. It is unsatisfactory that our youths know or understand the issue of environmental degradation, they have to do something at a personal level, engage the new ethic and mitigate the negative impact that is brought on by human action in the first place.

Finally, the Belgrade charter only provides a framework for developing school curricula to engage students in learning about environmental issues by outlining the aims of Environmental Education and by prescribing education objectives. It does not offer guidelines on the actual teaching and learning process. In short, it is only a four-page document that describes the key aims and objectives of Environmental Education that school curricula should consider. The purpose of this analysis is not to discredit a useful and important framework guiding Environmental Education curricula, but rather to explore ways in which a framework for CCE can be developed that not only provides guidance on curriculum but also offers more robust and effective ideas in the implementation of teaching and learning about the environment.

Despite these two well-discoursed frameworks of ESD and EE, we will first need to understand how climate change has been taught and learnt around the world before we can advance a more appropriate conceptual framework for CCE.

State of CCE around the world

Accurate knowledge about the causes of climate change is identified in several studies as a key determinant for informed action and the formation of an individual's personal resolve to counter the effects of climate change. Lack of knowledge and coherence on climate strategies led to powerlessness that resulted in inactive participation of pro-environmental behaviours (Kenis & Mathijs, 2012). Bord, O'Connor, and Fisher's (2000) research of the American public's knowledge about the global climate system shows that people who have inaccurate or erroneous understanding of the phenomenon could possess the same belief that global warming is occurring and may even have the ensuing intentions to act on such a belief. The implications are that while the study yielded results of low public understanding, it also explicates the fact that spurious knowledge does not affect actual behaviour change (e.g., no effect in voting for an environmental referendum) in a way that accurate knowledge does. To some extent, the results imply that behavioural change may be affected by simply raising awareness, regardless of how much understanding people have of the science of climate change. Indeed, when compared to participants from other regions of the globe, Hong Kong secondary pupils seemed to have a high degree of awareness of and positive attitudes about climate change concerns (Jackson & Pang, 2017).

On the other hand, other studies have shown that the awareness about climate change may not be translated into accurate understanding. In Canada, a cross-generational qualitative study by Pruneau et al. (2001) found that people may be able to indicate awareness about climate change without being able to identify its causes and consequences. Furthermore, research participants saw no tangible consequences of the phenomenon in their lives (Pruneau, Khattabi, & Demers, 2010). Similarly, a Yale University project aimed at determining US citizens' knowledge about climate change illustrates that the majority (63%) of Americans believe that global warming is happening, but many did not understand the reasons why it is so (Leiserowitz, Smith, & Marlon, 2011). The study also found important gaps in knowledge and misconceptions about climate change and the Earth system similar to those found in Chang and Pascua (2016). The need for functional public knowledge about climate change cannot be over-emphasised. The scientific community has shown increasing confidence in present and future threats as well as the extent to which humans contribute to the causes of climate change. It is therefore paramount that accurate information be transferred not to a few science experts, but to various stakeholders such as our students.

A similar study was also performed on American teens in middle and high school and the teens knew more or less than adults about how the climate system worked and the causes, consequences and solutions of climate change (Leiserowitz, Smith, & Marlon, 2010). American teens like their adult counterparts also have common misconceptions. In the same study, American adults and teens recognised their limited understanding of global warming and would like to know more about it. Other studies such as American beliefs and attitudes towards global warming have been conducted as well. One study focused on seventh-grade students' conceptions of global warming and climate change (Shepardson, Niyogi, Choi, & Charusombat, 2009). The study corroborates that students lacked a rich conceptualisation of the issue.

Studies of students' conceptions of global warming or climate change were replicated in other countries like Singapore, Iran, Turkey and Spain. Again, the study carried out in Turkey identified students' misconceptions of global warming (Kilinc, Stanisstreet, & Boyes, 2008). In addition, this study pointed out that many students believed in the scientific mechanism of global warming while concurrently believing in flawed explanations. In the Spanish study, the usual confusion with other environmental problems was detected (Punter, Ochando-Pardo, & Gracia, 2011). Students did not perceive the possibility of socio-economic problems and displayed certain confusions of ideas such as a lack of recycling will lead to an increase in climate change.

In Britain, the Climate Change Schools Project (CCSP) brought together organisations, schools and teachers to introduce CCE into the school curriculum. The project's objective was to develop a "Climate Change Lead School" network, advance an "adaptable and creative" climate change framework, and raise awareness and understanding of climate change, in terms of climate change teaching, learning and positive action (Grant & Featherstone, 2009). An evaluation report explored the effects on students' knowledge, attitudes and behaviours towards climate change and how students have influenced their peers and families. A survey was carried out with Year 5 and Year 8 students before and after participating in climate change lessons. Focus groups were also convened to explore what caused students to change or not to change their attitudes and behaviours towards climate change. At the end of this study, students demonstrated the main messages about climate change and the right message was brought across to them. Students also developed positive attitudes towards climate change and positive behaviours were observed (Grant & Featherstone, 2009).

This study has shown that even though students did not have a deep understanding of climate change and its impact, by introducing CCE, it helped bring about change to students' knowledge of climate change and their positive response to it in terms of attitudes and behaviours. Studies at the tertiary level like those of Salehi, Nejad, Mahmoudi, and Burkart (2016) also showed differences in climate change knowledge among students.

Educating and communicating on climate change is challenging. For instance, Hermans and Korhonen (2017) found that there is room for improvement in students' willingness to take action to mitigate climate change. As mentioned earlier, findings suggested there are climate change concepts that are often misunderstood. To bring a change to the mindset of people's attitudes and behaviour towards climate change, it is important to communicate the right message to the people. A meta-analysis of existing works to find out what the obstacles were uncovered cognitive, social, psychological and behavioural challenges that limited people's awareness, attitudes and behaviour actions (Pruneau, Khattabi, & Demers, 2010). Also, the article discussed some climate change communication and education strategies to remove such obstacles.

Again, a similar study was conducted in Australia on Year 6 students (equivalent to Singapore Primary 6). The findings indicated that after learning about global warming, students understood the causes and impact of climate change. The data also indicated an increased level of concern and "students' belief about their ability to make a positive impact in relation to global warming and climate change" (Taber & Taylor, 2009, p. 97).

Another research was carried out on how undergraduate students assessed the reliability of different information sources about climate change (Braten, Stromso, & Salmeron, 2011). Textbooks and official documents were reliable and trustworthy as compared to articles in newspapers or material published by a commercial agent. Students also ranked content over the date of publication when evaluating reliability. When assessing the reliability of a textbook, students were more stringent and emphasised different criteria than when judging other types of documents. The study also found that students who are less read in the topical knowledge tend to find less reliable sources trustworthy, unable to "differentiate between relevant and irrelevant criteria when judging the trustworthiness of sources" (Braten, Stromso, & Salmeron, 2011, p. 180).

Other studies have also explored how climate change should be taught in classrooms. In the article, "Content Literacy for the 21st Century: Excavation and Elevation and Relational Cosmopolitanism in the Classroom", the authors discussed two essential skills, excavation and elevation, needed by students to examine texts about climate change (Damico & Baidon, 2011). Next, the authors introduced a model, relational cosmopolitanism, for framing, understanding, learning and teaching climate change in middle school, secondary school and university level classrooms.

Another study was conducted on educated Americans about their beliefs and attitudes towards global change (Reynolds, Bostrom, Read, & Morgan, 2010). This study differed from the others as it was conducted at two different times, one in 1992 and the other in 2009. This study showed that after a decade long of attention to climate change, the survey respondents showed higher awareness and comprehension of some climate change causes. However, this study also pointed out that many individuals have incorrect beliefs about climate change (Reynolds, Bostrom, Read, & Morgan, 2010). The implications are that after paying attention to climate change or global warming issues over a period of time, people can be made aware of this issue.

One example was research on addressing moral aspects of climate change (Li & Cecilia, 2011). A study was conducted on 14-year-old students through group interviews and the data was analysed, taking into consideration both cognitive and situational aspects. The findings showed that the students conceptualised the solutions to mitigating climate change in relation to two different perspectives, either themselves or from another person's point of view. This is important as their views change accordingly to the different contexts and that students seem to be oblivious of their varying considerations for others, the environment and society.

While the literature on public education for climate change focuses on understanding and awareness, research on formal CCE extends the scope to curriculum design, instruction and assessment over and above attitudes and behaviour. In fact, one important aspect of CCE research is on teachers. If teachers are the key stakeholders in educating our young about climate change, then it is pertinent that we determine the teachers' readiness to deliver the CCE curriculum to our young.

In one study conducted on secondary school students and pre-service teachers in Australia, the subject knowledge, conceptual understanding and sources of information about climate change were examined (Boon, 2010). Results showed similarities between the two groups of participants.

This study pointed out the need to develop tertiary courses to improve student teachers' knowledge and gaps in important school curriculum topics, such as science and geography.

In another study, trainee teachers were subjected to a module on global climate change, and they were assessed on their knowledge, pedagogy, awareness of resources to address challenges with learners and ability to engage in meaningful conversation (Hestness, McGinnis, Riedinger, & Marbach-Ad, 2011). The findings suggested that through the inclusion of such modules, positive impacts were observed with regard to the points mentioned earlier. This can provide an insight into how teacher training educators and curriculum developers can help better prepare trainee teachers to deliver content to our young. Another study explored how schoolteachers can implement different components of CCE to bring across the right message and how to prepare students and strengthen their skills in critical thinking to face challenges posed by climate change in the future (Mckeown & Hopkins, 2010).

Certainly, we can see that there is room for more studies to be conducted with regard to CCE if we are to consider the role of education in mitigating climate change. What we can do is carry out more research in the context of CCE specifically and much more in-depth on the current state of CCE. This will establish baseline information that will shape recommendations on how climate change can play a more focal role in the school curriculum (Chua, 2010). Similar studies that have been conducted overseas might have been done in Singapore, but in the context of environmental education and relatively not in depth. Indeed, an in-depth understanding of CCE must consider the studies done at a national level and at the local level. In the next section, research related to CCE in the Singapore context will shed light on how the global understanding of CCE interacts with CCE in the national context.

State of CCE at the national and local scales

Case example of CCE at the national and local scales in Singapore

An early study on CCE by Goh, Tan, Chang, and Ooi (2009) discussed the state of CCE in Singapore by examining the formal school curriculum syllabus layout by the Ministry of Education (MOE). They also looked into the topics that were being taught and the gap that existed in formal school education. Informal CCE was also examined, on public or private driven activities that were carried out to raise awareness of climate change or global warming. This was done through document analyses. While there was not much research carried out to assess the public's knowledge, attitudes and behaviours about climate change in Singapore, two surveys carried out by the media focused on whether the public was aware of global warming or climate change (Mehlsen, 2009). In the survey conducted by HSBC, the public was assessed on their attitudes towards climate change (Chua's study,

2010, shows that Singaporeans are more concerned about the economy than climate change). However, in both surveys, they did not assess the public's level of understanding of climate change.

Prior to such studies, research that was carried out in Singapore was based on the general concept of environmental education, whereby environmental issues including climate change and air pollution were clustered together as one. For example, a study was conducted to assess the general environmental knowledge, attitudes and behaviour of secondary school and junior college students in Singapore (Tan, Lee, & Goh, 1998). This study provided insight into how Singaporeans think and what they believe in in terms of their environmental awareness. The results showed that Singaporeans were aware of environmental problems and keen to help in preserving the environment provided this was done within their comfort zones or if they perceive it to be beneficial. However, we need to note that this study was carried out in 1996, and there is a need to conduct a study to ascertain the current state.

There were also studies on environmental education in Singapore being carried out in recent years. Wee (2008) examined the state of environmental education in Singapore, discussing the MOE's approach to developing the curriculum and the limitations that are associated with it. A similar study uncovered three main factors that influence the development of environmental education in the curriculum, namely the pragmatic and utilitarian consideration of the urban state, high-stakes examination systems and the overriding motivation to balance between environment and development (Wong & Stimpson, 2003).

Another study was conducted by Nanyang Technological University, Wee Kim Wee School of Communication and Information, looking into how public campaigns can be designed to positively influence pro-environmental attitudes in youth (Chib, Chiew, Kumar, Lim, & Ale, 2009). Again, this study was conducted with a rather tangential treatment of environmental education, using recycling of plastic as an example. Even the most recent NCCS survey did not delve deep enough to assess the level of awareness and/or understanding of the public on climate change. Whether Singaporeans possess a significant basic knowledge about the phenomenon is not clear.

In terms of the school curriculum, Chang (2014) posited that the collaboration between agencies has resulted in the inclusion of varying weather and changing climates in the revised curriculum for 2014, as a way of introducing the climate change topic. The joint syllabus revision exercise with input from the National Environment Agency (NEA) highlighted environmental issues, particularly climate change to be featured more prominently at the secondary level (Pascua & Chang, 2015). Seen as a theme under Environmental Education (EE), it is interspersed in science and social studies with geography as the key school subject in examining the world climate system at the secondary level. Such topics were previously only taught at the upper secondary level (Chang, 2014) but were changed to include weather and climate changes at lower secondary levels as well. From lower to upper secondary levels, students are exposed to topics of climatic risks to Singapore such as temperature rise and coastal flooding (Curriculum Planning & Development Division, 2021).

Although CCE is found mostly at the secondary levels, it can be subtly found in other levels of education in Singapore. In primary schools, students are exposed to physical processes, environmental, scale and interdependence of geography in general science and social studies (Abalahin & Chang, 2020). Students at pre-university levels on the other hand are exposed to environmental topics in economics, social studies, science and geography (Chang & Pascua, 2017). The implementation of CCE across disciplines and educational levels indicate an acknowledgement of how environmental awareness should be cross-disciplinary and cultivated from an early age.

Further changes were also observed in the shift from traditional assessment modes to formative modes of learning. Yee (2020) noted that unlike past syllabi that were conceptual or systematically framed, an issues-based framework was adopted in the 2014 geographical syllabus, enhancing the potential for secondary geography education in Singapore. Through the implementation of "Geographical Investigations" (GI) in Singapore's geography curriculum, it acts as a "guided inquiry" process designed for learners to engage in criticality, reflective thinking and independent learning (Lim, 2018). Such shifts in the curriculum reflect meaningful development of CCE in Singapore that sees it adopting current pedagogical developments to engage learners in the learning process and to create opportunities for them to develop relevant skills to be applied in the real world.

Chang (2015) argued that the geography classroom is the best place to teach about climate change as it affords both space and time for students to interact with topics within the disciplinary domain of geographical subjects. Yet, Ho and Seow (2017) reported that the central focus in geography has been criticised for its traditional approach towards teaching environmental issues such as limiting the range of content knowledge taught and focusing more on the physical process. Reflected in Yee's (2020) study that analyses environmental knowledge (in different tiers) within the curricula, he described the incorporation of environmental knowledge in the curriculum in the first and second tiers that emphasises system causes, effects and strategies for change. Yee (2020) described that the third and fourth tiers, which are concerned with strategy effectiveness, alternatives and visions, are relatively weak.

In addition, Ho and Seow (2017) also noted the focus of the curricula is on educating students on the operations of environmental processes rather than developing an understanding of how societies work (Ho & Seow, 2017). Seow and Ho (2016) argued that social processes must be included within CCE for students to understand the bigger picture of how anthropogenic factors are caused by such structures. Even though students may learn about how rapid industrialisation in India and China has led to the release of greenhouse gases and how they affect climate change, they may not be taught about the economic, political and social structures these places have that make rapid industrialisation necessary (Ho & Seow, 2017). This results in a lack of understanding and awareness of the issue at hand.

Moreover, a comparative study between Singaporean and Philippine teachers revealed that Singaporean teachers adhered to the prescribed national geography curriculum, which focuses primarily on science and the impact of climate change, with a greater emphasis on "ecological literacy" and climate change scientific knowledge (Ho & Seow, 2017). In Singapore, teachers identify themselves within the geographical approach, and engaging in civic engagement and agency – which sees how other structures come into play – are not under their scope of teaching and are hence not taught (Ho & Seow, 2017). Instead, it is believed that civic engagement is under the social education approach which is found and taught in social studies instead. With social studies as another subject that incorporates CCE, it is safe to assume that other facets of environmental knowledge such as civic engagement and interrelations to social and political structures are taught within the curricula.

However, Chang and Pascua's (2017) study that analysed the curricular intent found a lack of content in subjects mentioned in the Strategy and Sustainable Blueprint in government policies aside from geography, social studies and science. Teachers are under the impression that there is an overlap in coverage across disciplines in climate change teaching, with non-geography teachers reporting on students entering the classrooms claiming to have learnt extensively on the topic (Chang & Pascua, 2018).

Accompanied by an over-burdened workload, non-geography teachers would rather skim and skip climate change topics that are not part of the syllabus and that are of no interest to students. Singaporean students are hence limited to factual and scientific knowledge of climate change. An added challenge is that not every student studies Geography subject at the upper secondary level. Upper secondary level geography – where students are able to learn more in-depth knowledge of climate change – is an elective subject for certain schools, indicating that only students who select geography as their humanity subject will gain more knowledge on climate change compared to their counterparts who did not select geography. Furthermore, CCE as a whole only exists in the pure geography subject at the secondary level where students learn about its entirety, such as its causes, impact and management of climate change (Chang, 2014).

In a study that investigated teacher perceptions, some teachers believed that students in regular government schools were not advanced or capable enough to deal with uncertainty and complexity (Seow & Ho, 2016). Discussion of controversies was seen to be more suited for academically more advanced students, feeling the need to minimise the amount of potentially confusing information for students who were academically weaker. Teachers, therefore, chose to manipulate facts and data to deliberately avoid introducing alternative perspectives so that students will not be confused (Seow & Ho, 2016).

Perhaps, the limitations in content knowledge of CCE play a role in the persistent misconceptions found in Singaporean students while learning about climate change. Using Chi's (2008) ideology on mental models, Chang and Pascua (2016) argued that concepts of climate change beliefs are organised into mental models that are not representative of the established geo-science knowledge of the phenomenon in students. The reasons for this observation include what Yang, Wei, and Su (2020) coin as "cognitive misers" where individuals are reluctant to spend cognitive efforts and instead rely on cognitive heuristics or mental shortcuts rather than actual knowledge as the perceptual filter to form opinions or make decisions. This results in students holding on to the erroneous perception of a coupling between climate change and other environmental issues such as pollution and global warming (Lambert, Lindgren, & Bleicher, 2012).

In attempts to diagnose and correct misconceptions, Chang and Pascua embarked on a few studies, such as an intervention-oriented evaluation study. The quasi-experiment used Patton's (2008) concept of intervention-oriented evaluation that posits as a process that is built-in and a time series to accommodate for midcourse changes. It was concluded that students showed significant development only after the teacher directly and explicitly refuted their responses compared to implicit refutations through contesting misconceptions (Pascua & Chang, 2015).

Their next study acted as a form of confirmatory research from the intervention-oriented evaluation, where data from the previous study was used to tabulate identified alternative conceptions in students which helped inform the construction of the test instrument. Certain tweaks, such as the absence of an assisted intervention were made to discount the researchers' own influence to allow teacher-led pedagogy to refute alternative conceptions (Chang & Pascua, 2015). Findings revealed a significant difference in scores from pre-test to post-test, providing further confirmation that several misconceptions were refuted successfully by instruction (Chang & Pascua, 2015). Resilience of misconceptions was once again noted in certain alternative conceptions being retained by students. The study concluded that unless direct refutation becomes the goal of instruction, alternative conceptions remain intact.

Utilising data from both research studies, Chang, Pascua, and Ess (2018) lastly aimed to correct student misconceptions with refutation texts at a local level. This is achieved by inducing conceptual change through a comprehension activity led by the teachers and supported by their peers, where a refutation-based instruction is implemented. Students were given two sets of refutation texts, one being a bogus set of text that contains misconceptions on climate change and an expert text that directly refutes misconceptions. Outcomes of the study highlighted a notable difference in scores for pre- and post-test, indicating statistically significant improvement in understanding in post-instruction (Chang, Pascua & Ess, 2018). This study was conducted *in-situ* within the context of one secondary school.

Another study by (Chang, Tan, Tan, & Kwek, 2020) investigated the use of learning progressions to correct misconceptions. Students were selected using a cross-sectional sampling scheme from the primary level to the university level. A hypothetical learning progression was developed from the Climate Change Diagnostic Test (CCDT) an instrument developed by Chang and Pascua (2016). The study found that most students possessed strong foundations at lower anchors but had persistent misconceptions at the upper anchors. Clear progressions were also witnessed from grades 5–11 but descaled upon reaching university levels. Such findings indicate various sources of confusion continually introduced into the students' learning journey from Primary to Tertiary levels and this goes towards the understanding where students did not know enough about the topic, and they were confused by things not explained in the textbooks.

Overall, these different studies that attempted to identify and correct misconceptions highlight the complexity and resilience of misconceptions, making it all the more necessary for knowledge of climate change to be comprehensive and not have gaps. A study on developing progressions for misconceptions revealed the negatives of centrally focusing on climate change at the secondary level. Additionally, teachers have mentioned how secondary students who study climate change to only pass their exams (Chang & Pascua, 2018) are only receptive to answers that are right or wrong (Chang, Tan, Tan, & Kwek, 2020). Accompanied by the lack of connecting CCE to broader societal structures and the limitations of content knowledge, all these promote a superficial engagement with CCE.

As mentioned earlier, education is critical in encouraging people to change their attitudes and behaviour. Implementation of CCE within the curricula is done so with the hopes to encourage youths to change their behavioural habits (UNESCO, 2021). However, such superficial engagement hinders students to reach their full potential of behavioural change. Posited by Chib, Chiew, Kumar, Lim, and Ale (2009), youth are seen as apathetic towards environmental issues despite an NEA poll revealing 90% of youth awareness of environmental issues. According to the global competence segment of the Program for International Student Assessment (Pisa), 8 out of 10 students in Singapore know more about issues like climate change and global warming, higher than the global average of about 6–7 out of 10 students (Teng, 2020). Despite this higher level of awareness, plastic recycling between 2019 and 2020 was found to have stayed stagnant at 4% (NEA, 2020). It is clear that a gap in knowledge and behavioural attitudes exists and persists among students.

As we can see from the examples above, there is plenty of room for further research on CCE in Singapore. If education is to be employed as an effective mitigation strategy for climate change, more research has to be done in this area to appreciate the existing understanding of the target population so that efficacious programmes can be developed to engage the populace. For example, similar research like the one conducted in 1996 by Tan, Lee, and Goh

(1998) can be repeated to assess people's knowledge, attitudes and behaviour, but with the topic changed to climate change or global warming to reflect the current situation.

Climate change education among the public

According to Chang and Pascua's (2018) teacher perception study, teachers posit a disjoint in concept-based classroom learning, where values learnt in schools are not reinforced when students step out of the classroom, calling to the inconsistencies of practices by the general public and what is taught in school. Despite encouraging children to practice recycling this can be deterred by the unwillingness of their family members to follow suit. From this, we see that the gap in knowledge and behavioural attitudes in students is attributed to family behavioural habits at home as well. Such findings reveal a lack of effort in engaging the public in cultivating environmental habits.

Ho and Chuah (2017) reported that climate change communication is effective and extensive through the 3Ps: the Public (government sector), the People (non-profit organisations) and the Private sector. At that time, there was an increasing portrayal of climate change and environmental knowledge through media and activities in Singapore (Ho & Chuah, 2017). Even though the public was made aware of environmental issues through campaigns and programs, pro-environmental attitudes and behaviour were still found to be getting worse. Reflected in Ho and Chuah (2017)'s study, a survey conducted on the public's perception of climate change illustrated a decrease in individuals believing it is a cause for concern and their responsibility to deal with climate change. A previous survey done by the NCCS a few years ago depicted a totally different story: where citizens were concerned about climate change and felt that it was their responsibility to do something about it. In addition, domestic plastic recycling rates from 2016 to 2020 fell from 7% to 4% in 2018 and continued to stay stagnant until 2020 (National Environment Agency, 2020). The lack of successes in making the community adopt pro-environmental action in spite of proper communication channels marks a stark indication that the knowledge and education programs put out by the government are not working.

Wi (2019) claims that these failures in the public sphere can be attributed to the polarity between policymakers and citizens. Information flow was stated to be unilateral and not tailored to the interest of citizens, with many feeling that the information presented was dry, uninformative and impractical to them (Wi, 2019). Furthermore, climate change campaigns approved and developed by policymakers have been guided by an information deficit model that assumes that the lack of knowledge hinders individual behavioural change. The importance of increasing an individual's factual knowledge is stressed upon, insisting that enhanced public understanding of climate change will result in attitude or behavioural change. However, as illustrated by many literature studies, a weak relationship between knowledge and behavioural change exists, with studies examining the role of factual knowledge in behavioural change yielding little effects (Yang, Wei & Su, 2020). Similar to CCE in the education, focus on factual knowledge further reflects an oversimplification of CCE and an expert lay divide in the community.

Even though weak links are revealed between knowledge and pro-environmental action, it does not necessarily mean that education plays no role in helping the individual achieve behavioural change. What is required is for individuals to find meaning and value in the knowledge that is offered to them (Wi & Chang, 2019). Factual knowledge that is provided to citizens do not give them meanings as to why individuals should change their habits. For example, an individual may receive the knowledge of setting the thermostat of an air-condition higher but does not understand "why" it is needed to do so. As such, Wi and Chang (2019) argue for a case of transformative education, where learners are able to transform information into knowledge, ultimately creating action based on the new knowledge acquired. These types of education are different from the transmission of factual knowledge that does not explain or help individuals understand given information. In this case, transformative education not only gives factual knowledge but helps explain the necessity to do so.

To test if transformative education works, a Transformative Education for Climate Change (TrEC) program was established and implemented in a residential area, lasting three months. Empirical findings showed that the program was successful, with improvements in both participants' knowledge and behaviour in the form of performing pro-environmental action (Wi & Chang, 2019). Despite a successful implementation, there were still a small number of residents that did not feel inclined to change their behavioural habits. These reasons were attributed to the overriding egoistic and hedonistic values of inconvenience and pragmatism, with certain individuals providing more insight by stating that they will only participate in pro-environmental action if the government provided incentives such as monetary benefits to make them more motivated (Wi & Chang, 2019). Such findings illustrate how individual values contribute to motivations in changing behaviours and would therefore be worth considering for CCE to target transforming individual values and beliefs to result in pro-environmental action.

Additionally, a percentage increase of individuals was found to believe that it is the government's responsibility to combat climate change (Ho & Chuah, 2017). This is similarly witnessed in Wi & Chang's (2019) study that revealed how citizens felt that the government was responsible for the issue of climate change. Posited by Detenber, Rosenthal, Liao, and Ho (2016), the government has been so effective in managing the city-state's natural resources and infrastructure that has created the impression among many citizens that it can solve most problems and that individuals need not act themselves. This results in an overreliance on government action which is no exception to climate change, especially when individuals perceive these issues to be something that they cannot control. This illustrates how government responsibility can be further attributed to the gap between policymakers and citizens mentioned earlier.

Wi (2019) asserts that grassroots organisations and residential communities should be used to better engage the community and get people to act and behave in ways that are beneficial to the environment. Working closer to the ground, these organisations and residential committees are not only able to understand the requirements and concerns of the neighbourhood, but act as a bridged communication between the government and residential estates. By doing this, both a top-down and a bottom-up approach come together to help people communicate with each other better and turn their ideas into actions. Although such an approach is beneficial for both communities and the government, Wi (2019) highlights the need for different initiatives based on different demographic patterns as well.

Demographic differences in climate change knowledge distribution were found all across Singapore (Yang, Wei, & Su, 2020). Adopting a nationally representative door-to-door survey in Singapore, Yang Wei, and Su (2020) investigated how demographic differences played a role in climate change knowledge and found significant results. In each demographic (gender, age, education and economic background) type, significant differences were found in climate change knowledge. It was discovered that males, young adults and individuals with a higher educational background or economic background were more knowledgeable about climate change in comparison to females, middle-aged and elderly individuals and individuals with a lower educational background or economic background (Yang, Wei, & Su, 2020). Such findings yield practical implications for designing climate change campaigns and educational programs, suggesting that people from different social groups tend to have different levels of actual knowledge. Illusory knowledge was also found to be different in distribution according to demographic groups (Yang, Wei, & Su, 2020). As such, education programs should be specialised for specific target audiences due to different understandings of climate change. With the quality of life being different between these demographic groups, it is also safe to assume that value and belief systems are different for these demographic groups as well.

Conceptualising climate change education

Instead of deriving a new domain of study that we brand as CCE, this book adopts the conceptual framework of environmental education to discuss how people can learn about the causes, impact and management issues of climate change. This also takes into consideration the findings from studies in international, national and local contexts.

When we say someone has learnt something about climate change, we need to be able to describe what the person has learnt. Do we mean that the person has learnt some facts about climate change or do we mean if the person has understood a concept? In short, what sort of learning outcome can we observe if a person is said to have learnt something about climate change?

Historically, educators within a behavioural paradigm considered knowledge to be stored as part of the learner (or his memory) when something is learnt. The theories regarding learning conditions, however, have evolved to incorporate cognitivist psychology theories later, specifically the information-processing model of cognition in which the conditions of the learner's environment influence human performance as internal processes can be influenced by external conditions (Gagné, 1985; Gagné & Medsker, 1996). When thinking about the conditions that are required for some capabilities to be learnt, Gagne suggested that it is not simply naming what is to be learnt but the learning capabilities that make students capable of accomplishing things they could not previously carry out. It is these capabilities that make up the outcomes of learning. Gagné (1985) advocated that learning does not merely involve recalling the name of what has to be learnt but rather allows the learners to accomplish new things. It is these competencies that make up the outcomes of learning. For instance, when we say that a student has learnt about the "greenhouse effect", we mean that the student might be learning how to define the greenhouse effect, what are the inputs and outputs to the atmospheric energy balance, differences between the natural and enhanced greenhouse effect or explain the various processes of heat absorption and re-radiation.

When Gagné refers to learning outcomes, he really is referring to the desired capabilities of a lesson or performance categories. These performance categories indicate conditions most favourable for the type of learning outcome observed. This concept is similar to that of instructional objectives as proposed in *The Taxonomy of Educational Objectives, Handbook I: Cognitive Domain* (Bloom, 1956) which presented a classification of subject matter or how content might be processed, aimed at helping curriculum builders "specify objectives so that it becomes easier to plan learning experiences and prepare evaluation devices" (Bloom, 1956, p. 2). Implicit in this definition is that there are specific outcomes of learning that can be classified and used as specific instructional objectives for curriculum planning. However, it was Gagné who explicitly proposed levels of learning outcome. In the modified and improved Gagné and Driscoll (1988) system, cognitive, affective and psychomotor domains were incorporated into differentiating learning outcomes.

Over the past five decades, variants of Bloom's and Gagné's classification have been used for understanding learning outcomes. Krathwohl's (2002) critique and revision of Bloom's original taxonomy were based on how the original instructional objectives were constructed. He noted that the statements that describe the intended learning outcomes were based both on the subject matter as well as a description of what is to be done with this subject matter. Consequently, these outcomes were expressed both as nouns or noun phrases with a verb or verb phrase, indicating both the subject matter content and the cognitive processes. For example, a statement such as "The students will be able to remember the definition of greenhouse effect" is framed around the noun phrase "definition of greenhouse effect" and the verbal phrase "to remember", reflecting both the content knowledge and cognitive process. Krathwohl suggested separating the two and articulating clearly the outcomes in the knowledge dimension as well as the cognitive

Table 2.1 Krathwohl's two-dimensional taxonomy

The knowledge dimension	*1. Remember*	*2. Understand*	*3. Apply*	*4. Analyse*	*5. Evaluate*	*6. Create*
A. Factual Knowledge						
B. Conceptual Knowledge						
C. Procedural Knowledge						
D. Metacognitive Knowledge						

Source: (Krathwohl, 2002, p. 216).

dimension. The knowledge dimension comprises factual knowledge, conceptual knowledge, procedural knowledge and metacognitive knowledge. The cognitive dimension includes remembering, understanding, applying, analysing, evaluating and creating. Each of these has further classifications (Table 2.1).

To adapt this classification for the understanding of climate change learning outcomes, I propose the following table (Table 2.2).

Nevertheless, there are still some overlaps in the categories within the knowledge dimension. For instance, the application of a piece of factual knowledge can be argued to be the recall of procedural knowledge. Moreover, in using the example of a student learning about the role of the enhanced greenhouse effect in causing global warming. A student may interpret that

Table 2.2 Proposed two-dimensional taxonomy of learning outcomes for CCE

The knowledge dimension	*1. Recall*	*2. Understand*	*3. Apply*	*4. Analyse*	*5. Evaluate*	*6. Create*
A. Factual Knowledge						
B. Concepts						
C. Skills						
D. Values/ Attitudes						
E. Actions						

Source: (Krathwohl, 2002, p. 215).

increases in CO_2 emissions over the last 150 years have coincided with a global average temperature increase of 0.8°C and is likely the cause of the temperature increase. This could be because that student has applied what he has learnt using the factual information, but he may also be recalling the relationship from something he has read on the Internet. To this end, the two-dimensional classification of learning outcomes presents difficulties, at least in its classification of the knowledge dimension.

Moreover, this two-dimensional approach of examining learning outcome is helpful, in that it clarifies the dimension of the proposed taxonomy. However, climate change knowledge is not a simple body of knowledge, in that depending on whether a student learns about the causes, impacts and management of climate change, different concepts and skills are required. I, therefore, propose that in order to put the classification of what students have learnt about climate change into perspective, we must distinguish if the knowledge gained is with respect to the causes, impacts and management of climate change. The categories are reorganised in Figure 2.1 to illustrate the proposed taxonomy.

The proposed taxonomy of learning outcomes provides a vocabulary that can describe what a student has learnt about climate change. In other words, we have a matrix with which to describe and discern the extent to which education on climate change has provided the necessary conditions for developing a better-informed citizenry that can better manage climate change issues in the future. To illustrate, we can describe the learning outcomes of any particular event planned for raising climate change awareness using this framework – whether it has simply resulted in remembering to turn off the light before leaving the room or the ability to analyse and critique the latest IPCC report.

However, this matrix of domains remains difficult for visual comprehension and an attempt was made to derive a visual representation of the framework that would work. To do so, a community of teaching professionals was gathered as part of a professional development workshop that focused on defining CCE, discoursing a framework for CCE and discussing issues of curriculum, instruction and assessment for CCE. The participants were provided with a similar discourse presented in this chapter and were encouraged to represent their understanding in graphical formats through concept mapping. Figures 2.2 and 2.3 show a few examples of the graphical representation that were produced by the teachers.

While the graphical representations presented above were an abstraction of Figure 2.1, consequently making it easier to understand the framework, Figure 2.4 provides the best representation.

In the original framework, the conceptions are generally sequential but in the graphical representation in Figure 2.4, there is an incorporation of all three dimensions of learning dimensions for climate change. The entire framework is circumscribed by the context of the subject matter pertaining to climate change causes, impact and management. The subject matter content can then be described by the domains of cognitive engagement, namely

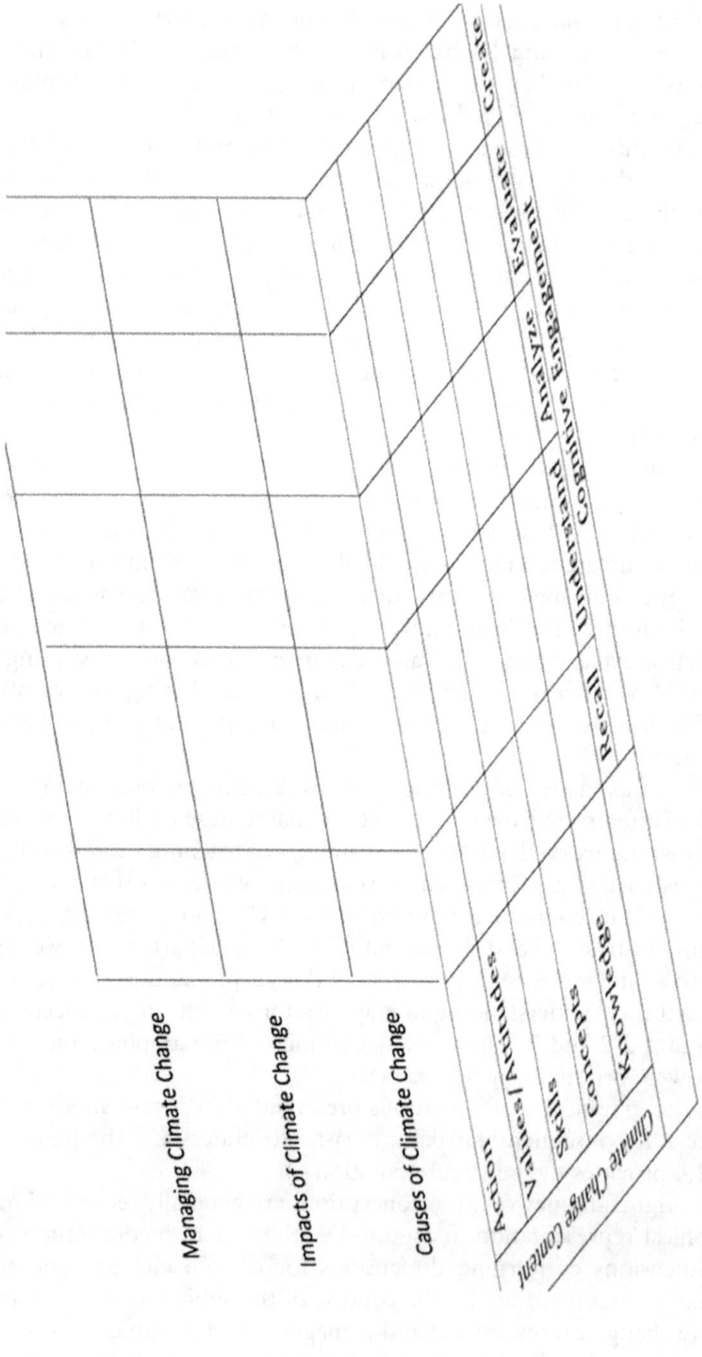

Figure 2.1 Learning outcomes in climate change education.

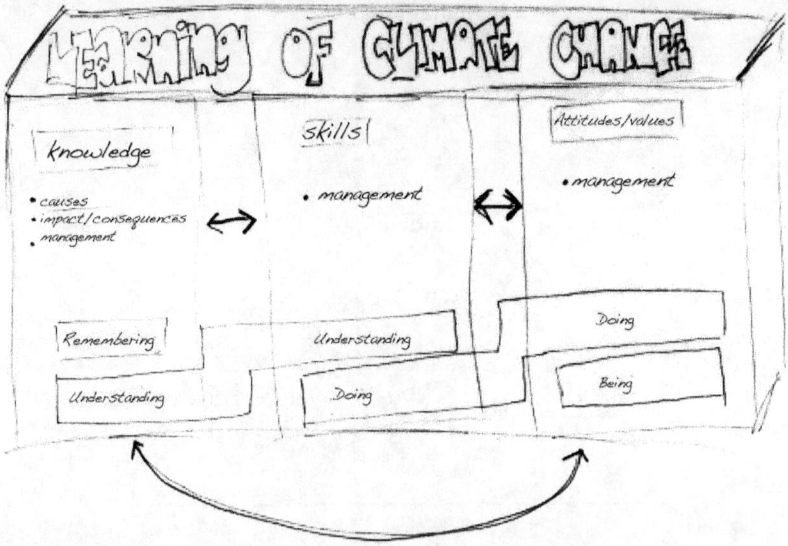

Figure 2.2 Example of a graphical representation of the conceptual framework.

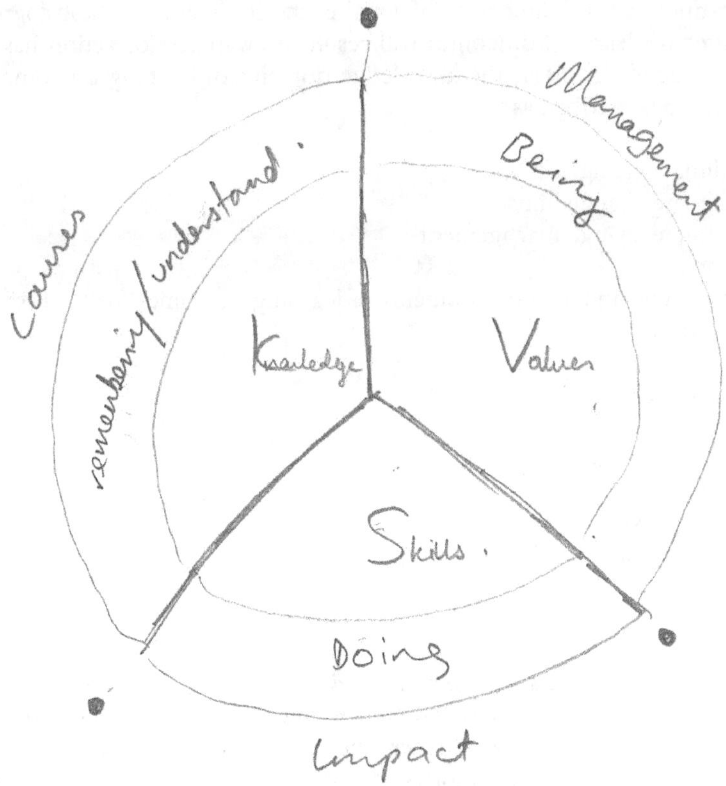

Figure 2.3 Another example of a graphical representation of the conceptual framework.

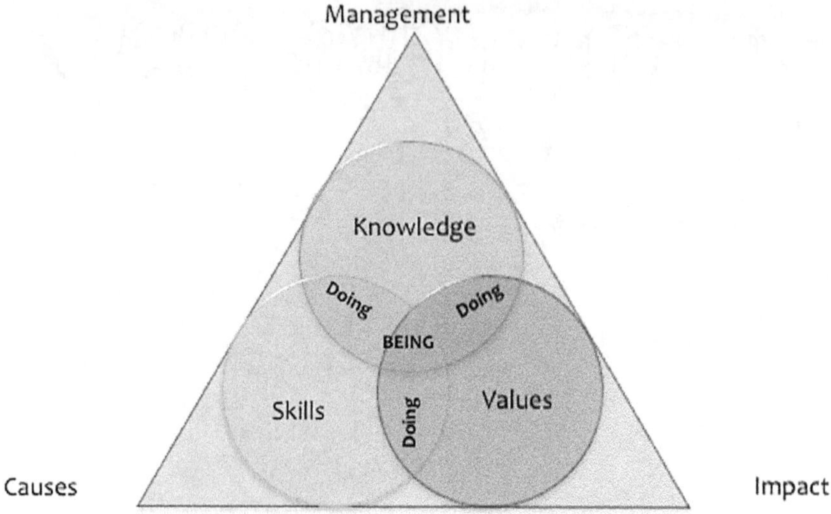

Figure 2.4 Conceptual framework of learning outcomes in climate change education.

knowledge and skills and the affective domain of learning – values. Interestingly, a third dimension of whether the confluence of these cognitive and affective learning outcomes will result in awareness or action has also been included. In short, the knowledge domains of learning outcomes for CCE can be described as

I. Climate change causes
II. Climate change impact
III. Climate change management

The cognitive and affective domains of learning outcomes for CCE can be described as

1. Knowledge
   a. Recall
   b. Understand
2. Skills
   a. Analyse
   b. Evaluate
3. Values
4. Action

In summary, *climate change education seeks to help learners develop knowledge, skills, values and action to engage and learn about the causes, impact and management of climate change*. Further, the dimensions can be collapsed and illustrated in Table 2.3. Exemplars of what each category means have been included in the table.

Table 2.3 Graphic organiser of learning outcomes for CCE (with exemplars)

	Causes of climate change	*Impact of climate change*	*Managing climate change*
Knowledge • Recall	Able to name greenhouse gases	Able to state that sea level will rise	Able to recall that the thermostat should be tuned to 25°C
• Understand	Able to explain the enhanced greenhouse effect	Able to explain how the thermal expansion of water will result in sea level rise	Able to explain how electricity production through fossil fuel combus-tion contrib-utes to the greenhouse effect
Skills • Analyse	Able to read graphs and determine the correlation between GHG concentration and temperature trends	aAble to use information sources such as hazards maps to determine the social-economic impact of a storm	Able to compare the pros and cons of raising land levels to combat sea level rise
• Evaluate	Able to use multiple sources and argue for or against human-induced climate change	Able to question and determine if the IPCC report takes land ice melting at the poles into account in projecting sea level rise	Able to evaluate the effective-ness of adaption strategies such as raising land levels in combating sea level rise
Values	Makes informed judgement on a stand for or against human-in-duced climate change	Makes informed judgement that storm frequency has increased due to climate change	Believes that the individual can do something to combat climate change
Action	Convinces others of his or her own belief about the causes of climate change	Convinces others of his or her own belief about the impact of climate change	Changes personal habits to reduce carbon footprint

Although the framework can be described as old wine in new bottles, in that the tenets and principles for teaching climate change do not detract from those of ESD and EE, the organising concepts make understanding this framework more readily understood. This framework addresses the key learning outcome domains for Environmental Education, from awareness to knowledge and ultimately to participation. Further, the framework addresses global and national issues of environment, society, culture and economy, is locally relevant and culturally appropriate, addresses content and builds civil capacity for environmental stewardship and a good quality of life consistent with the guidelines for ESD. In truth, the CCE framework developed in this book is not meant to replace ESD and EE but rather to customise a framework that supports the teaching and learning of climate change specifically.

Summary

CCE's goal is to educate and inspire the next generation of environmental stewards by fostering critical thinking and participation in the dissemination of new knowledge. Teaching critical thinking skills to students is crucial for the survival of the planet in an era of "fake news" and deceit. For a global citizen, what does CCE mean? Climate change's local effects provide a challenge for CCE because students must reconcile the changes they perceive on a local or national level with the global impact they read about. This chapter introduced a conceptual framework for CCE discussions, taking into consideration the climate change effects across global, national and local scales. The chapter offers a framework against which we can design a curriculum that meets the requirements of our students through an examination of CCE in global, national and local contexts.

In order for educators to implement this framework to interpret and enact a CCE curriculum, having a macro-framework to organise the teachers' subject matter knowledge is necessary but not sufficient. There is a need to map the relevant topics from syllabus documents across subjects and levels. The following chapter (Chapter 3) will examine the place of CCE in the formal curriculum, using the Singapore case as an example.

3 Climate change education in the formal curriculum

Climate change education (CCE) is an important strategy for combating the effects of climate change. The United Nations, through UNESCO (2020), reaffirmed the essential role of education in raising climate change awareness, enhancing adaptation capacity and reducing vulnerability to climate-related disasters in its most recent Global Education Monitoring (GEM) report. It emphasised the necessity for a deliberate action of fostering climate literacy for teachers, students and communities, and called for a holistic participation of the education sector in the issue. In particular, CCE can be incorporated into the school curriculum from as early an age as possible.

The governance of CCE is based on the United Nations Framework Convention on Climate Change's (UNFCCC) call for governments to establish and implement education programs to prepare society for the threats and opportunities posed by a changing climate. For the purposes of this chapter, CCE also refers to Climate Change Education for Sustainable Development (CCESD) because it is a flagship initiative of UNESCO's (2010) second half of the UN Decade for Education on Sustainable Development (DESD).

CCE is an important strategy for combating the effects of climate change (Anderson, 2010; Bangay & Blum, 2010; Leal Filho, Pace, & Manolas, 2010), which has been dubbed the "defining challenge of our time" (UNESCO, 2010, p. 2). Education is considered a tool for communities and individuals to make informed decisions about how to address the challenge of climate change and to improve their adaptive skills in dealing with its approaching repercussions. While some countries have been hesitant to include CCE in their climate change agenda (Kastrenakes, 2014), many others have (Læssøe & Mochizuki, 2015), with the school curriculum providing a suitable structure for knowledge distribution and skill development.

UNESCO (2015) published case studies of different nations' experiences with CCESD in *Not Just Hot Air: Putting Climate Change Education into Practice*. The report highlighted inconsistencies in policy articulation on CCE, with some nations failing to include the topic in national education strategies while others have emphasised including the tropic in the formal curriculum. Such policy directives are usually issued from top to down, involving inter-agency collaboration and are part of a larger philosophical and practical approach to climate change management.

DOI: 10.4324/9781003093800-3

CCE in the formal curriculum

In a cross-national examination of 17 nations' policy texts on CCE, Læssøe and Mochizuki (2015) found that states use "soft" governance rather than legislation when deploying education for climate action. While the value of education is recognised, it remains an underutilised resource embedded within incoherent policy frameworks (Mochizuki & Bryan, 2015), occupying a marginal slice of the pie in national policymaking and public administration (Han, 2015; Læssøe and Mochizuki, 2015) and, at times, subject to the policy whims of elected governments as they come and go (Sung, 2015). Indeed, even where robust climate regulations are in place, such as in Brazil, legislation that mandates education as a powerful tool for climate action is lacking (Trajber & Mochizuki, 2015). In the absence of state-mandated regulatory mechanisms, Læssøe and Mochizuki (2015) emphasised the importance of "norm supporting structures" that will advance the discourse and raise stakeholders' educational awareness.

Indeed, the UNESCO along with the UNEP stress the strategic role of education in combating climate change in both the areas of mitigation and adaptation (UNESCO & UNEP, 2011). The two organisations maintain that policies and curricula should be synergised with the common goal of securing knowledge foundations with an emphasis on deepening understanding of causes and impacts, promotion of skills, values and attitude development "using appropriate action-oriented pedagogies" (UNESCO & UNEP, 2011, p. 55). However, it has been the trend that climate change, particularly, topics on global warming and the greenhouse effect are integrated into existing school subjects. In Greece, for instance, the greenhouse effect is not explicitly included in the curriculum but related concepts and ideas are integrated into textbooks content (Koulaidis & Christidou, 1999). The coverage and depth of discussions are patchy at best, leading to a lack of rich conceptualisation in the understanding of climate change (Shepardson, Niyogi, Choi, & Charusombat, 2009).

CCE in the school curriculum in Singapore

McKeown & Hopkins (UNESCO & UNEP, 2011) posit that CCE as tackled in the formal setting tends to be contained in the conventional view of teaching atmospheric composition following a natural science perspective. This is indeed the case in Singapore's formal curricula context (Chang, 2012a). The discourse then follows a rigid structure, devoid of the robust and interdisciplinary approach that the authors espouse. Further, they suggest that the best way to harness the potential of education, as a resource is to explore opportunities that involve both formal and informal education means.

This chapter will present how climate change issues and related concepts are being dealt with in Singapore's current curriculum. The intention for using the Singapore case is deliberate so that the local case can be compared to studies around the world, with a view to inform the teaching and learning of climate change in a wider context. Within the formal curriculum, school textbooks from the primary level to the secondary level were analysed for their contents in terms of extent and scope. This provides insight as to how CCE is integrated into the different disciplines of the natural sciences, social studies, the humanities and moral education.

Up until the year 2020, the Singapore government has a well-defined climate change plan but no clear statement on how CCE might be implemented in formal education can be found. While climate change is a topic taught in schools, mostly through geography as a subject for environmental education, it is only mentioned in national policy papers as being embedded in the school curriculum. CCE appears to be performed in the lack of conformance to directions from top-level state entities in a city-state recognised for its highly systematised operations. Chang and Pascua (2017) found that Singapore's national-level climate change policy has not provided a direct mandate on how CCE should be carried out, but there is a clear indication of the presence of CCE in the curriculum. They argue that this coincidence of the coverage of the topic in "curriculum with the intent of the state's climate change policy" was largely supported by the strong professionalism among teachers and curriculum developers. Since then, the 2021 Green Plan has explicitly defined an Eco-Stewardship programme among other initiatives to tackle the problem of climate change and the alignment between the state-level policies and the curriculum has become very apparent.

Formal curriculum in Singapore

Singapore children attend six years of primary or elementary school and four years of secondary education (five-year secondary course for slower learners). National level examinations are held at the end of primary (Primary School Leaving Certificate) and Cambridge O-level at the end of secondary. Beyond secondary education, students can study a two-year academically oriented junior college education culminating in the Cambridge Advanced Level (A-level) examinations or a three-year polytechnic diploma at any of the five polytechnics before they apply for admission to universities (Ministry of Education, Singapore, 2012) (Figure 3.1).

Primary education is mandatory in Singapore under the Compulsory Education Act 2000 (Ministry of Education, 2022). Out of those who enrol in Primary 1, some 39.1% will finally go to a university and end up with a degree, 49.2% to polytechnics ending up with a diploma, another 25.4% to the institutes of technical education while the rest enter the job market after secondary school (Education Statistics Digest, 2022).

The literacy rate for those above 15 years old is 97.1% (Department of Statistics Singapore, 2012). Enrolment in all educational institutions totalled 654,937 in 2019 with primary schools accounting for 235,039, secondary 161,831, polytechnics 87,089 and universities 107,630 (Department of Statistics, 2022b). There are 28,330 teachers in some 316 primary and

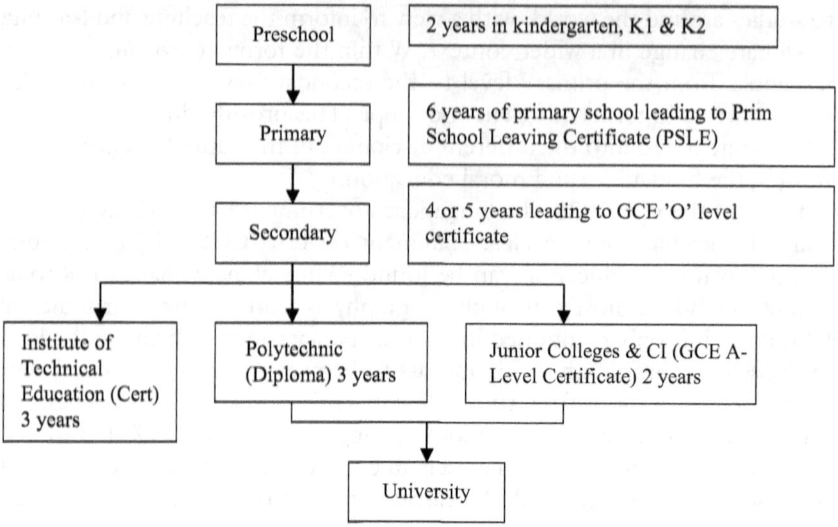

Figure 3.1 Simplified illustration of Singapore's schooling system.

secondary educational institutions in Singapore (MOE, 2022). Education is generally pervasive and ubiquitous.

For the purposes of using a comparable reference system to the levels of study across the Singapore and international education contexts, grades 1–6, 7–8, 9–10 and 11–12, will be referred to in this book as primary, lower secondary, upper secondary, pre-university levels, respectively. Table 3.1 provides a more visual representation of this system of reference.

In Singapore, Environmental Education at both the primary and secondary levels is generally integrated into science and social science curricula, as well as in civics and moral education. Students' acquisition of knowledge and understanding of the environment and in developing a sense of responsibility for the environment are emphasised (Curriculum Planning and Development Division, 2007a). A summary of the place of CCE in the formal school curriculum is included in Table 3.5 for reference and discussion (Table 3.2).

While social studies and civic and moral education are subjects studied by all students in primary schools, CCE focuses on the child taking action for their environment. However, there is little knowledge and skills taught

Table 3.1 Grade equivalence of levels of study in Singapore schools

Level of study in Singapore		*Grade level equivalence*
Primary 1–6	Primary	Grades 1–6
Secondary 1–2	Lower Secondary	Grades 7–8
Secondary 3–4/5*	Upper Secondary	Grades 9–10
Pre-University 1–2	Pre-University	Grades 11–12

* Note that even in a five-year secondary programme, the scope of learning at the end of Secondary 5 is of roughly similar scope to Grade 10 in the American system.

Table 3.2 Place of CCE in the primary school curriculum

Level of Study	*Subject*	*Topic(s)*	*Comments*
Primary 1	Social Studies	Knowing Myself, Others and My Surroundings	The issue of Climate Change is not explicitly spelt out in the syllabus document. This section will introduce students to their connection to the environment and how their actions can affect the place around them
Primary 3	Social Studies	Understanding Singapore's Environment & Challenges	"[L]earning about the consequences of their actions on the environment and importance of resources" (Curriculum Planning & Development Division, 2020, p. 16)
Primary 4, 5 & 6	Civics and Moral Education	Protecting the environment	Students learn that to continue living in a healthy environment, they need to protect the environment, as the damage is irreversible. They are encouraged to examine the products that they use in their everyday lives and consider if these have adverse effects on the environment, find alternatives if they do and be involved in recycling activities. This subject is taught by a mother tongue teacher (Chinese, Malay and Tamil) who may not have the geography or science background to understand the issues (Curriculum Planning & Development Division, 2007)
Primary 4, 5 & 6	Science	Cycles in Plants & Animals Cycles in Matter & Water	Students are to develop an inquisitive nature towards animals and plants, to instil care and concern towards nature. Concern for water is also inculcated to understand it as a limited resource and its need for conservation (Curriculum Planning & Development Division, 2013)

Summarised from (Goh, Tan, Chang, & Ooi, 2009) and updated based on syllabus documents from 2007, 2013 and 2020.

at that level. In the science subject, students learn more about nature and water resources specifically but not directly about the climate change topic. The concepts introduced at the primary level are revisited in greater breadth and depth in the lower secondary curricula. Apart from the direct topics of climate change, students learn about "showing care and concern for living things and awareness of our responsibility for the quality of the environment" (Curriculum Planning and Development Division, 2007a, p. 9) (Table 3.3).

Table 3.3 Place of CCE in the lower secondary school curriculum

Level of study	*Subject*	*Topic(s)*	*Comments*
Secondary 1 & 2	Geography	Introduction to Geography	Students learn about how they can sustainably manage natural resources. Here, students are exposed to "different views of natural resources" and how a human-centred view "can lead to over-extraction and subsequent environmental challenges" (Curriculum Planning and Development Division, 2021a, p. 9)
Secondary 1 & 2	Science	Interactions Transfer of Heat Energy and its Effects	Students learn how "interactions with our environment" through the transfer of energy can cause changes in the environment that may "intensify natural disasters and cause chemical changes which can affect the ecosystem". This will help students "appreciate how Singapore is adapting to climate change and how as individuals can do their part to mitigate climate change" (Curriculum Planning & Development Division, 2020, p. 42)
		Interactions within Ecosystems	Students will "learn how climate change can affect ecosystems and what they can do to maintain biodiversity" (Curriculum Planning & Development Division, 2021b, p. 48). Explore how sustainable living is practised through their interactions with the environment
		Exploring Diversity of Matter by their Physical Properties	The topic here enables students to explore main classes of materials for students to "display habit of sustainable use of materials and practice the 3Rs". Students will also get to "evaluate the environmental impact of excessive use of paper & disposal of plastics" (Curriculum Planning & Development Division, 2021b, p. 32)

Summarised based on syllabus documents from 2020 and 2021.

Table 3.4 Place of CCE in the upper secondary school curriculum

Level of study	*Subject*	*Topic(s)*	*Comments*
Secondary 3 & 4	Social Studies	Being part of a globalised world	Students learn to empathise and care for the environment that they live in (Curriculum Planning & Development Division, 2016a, pp. 8, 10)
Secondary 3 & 4	Geography	Variable Weather & Changing Climate – A continuing challenge?	In the topic on varying weather and changing climate, the knowledge and history about the causes, impact, patterns and management of climate change related phenomena are included (Curriculum Planning & Development Division, 2017)

Summarised from (Goh, Tan, Chang, & Ooi, 2009) and updated based on syllabus documents in 2016 and 2017.

The topic of climate change is not taught as a stand-alone topic in the lower secondary subjects but is included as part of different topics such as heat transfer or ecosystem diversity in the science subject. However, there is always the contextualisation of how the topics they learn have implications for the environment and, in particular, on climate change in some instances (Table 3.4).

Geography is the main subject that CCE is present in the subject matter knowledge. Although there are components of caring for the environment in the Social Studies subject, they are contextualised within the discussion of how a student is part of a globalised world. In Chang and Pascua (2017) and Chang, Pascua and Ess (2018), the focus of those studies was on the Secondary 3 and 4 Geography subject which discusses the subject matter that is covered for the topic. There is a further elaboration of those two studies in the subsequent chapters (Table 3.5).

Students in Singapore have to prepare for the Singapore-Cambridge GCE A-Levels examination as the summative assessment of their course of study at the pre-university level. Pre-university subjects in Singapore are organised into three levels, Higher 1, Higher 2 and Higher 3 in order of increasing breadth in covering the subject matter knowledge. An H2 Chemistry subject, for example, would cover more topics than an H1 Chemistry subject, but the difficulty level of the questions answered would be nearly the same. The only direct mention of the climate change topic at the pre-university level subjects is in H1 Geography and H1 Biology while the other aspects of climate change impact are discussed through the topic of sustainable development.

As the combination of subjects is similar for all students up to the lower secondary level, all students are exposed to the same knowledge base of CCE. However, subjects such as social studies and moral education are

Table 3.5 Place of CCE in the pre-university school curriculum

Level of study	*Subject*	*Topic(s)*	*Comments*
Pre-University (H1)	Geography	Climate Change Urban Change	The section is organised around the relationship between climate change and extreme weather patterns through the study of floods and the tropics. Specifically, the section introduces knowledge on CCE such as the effects of climate change and our responses to them. Sustainable development is also introduced (Curriculum Planning & Development Division, 2016b)
Pre-University (H2)	Geography	Sustainable Development	Sustainable development is the main concept in the unit. In terms of skills, students are required to "understand how environmental management is necessary to ensure economic growth" (Curriculum Planning & Division, 2016c, p. 8)
Pre-University (H1)	Biology	Impact of Climate Change (Extension Topic)	Students will explore how "way of life influences climate change" and how climate change links to the evolving "threat of infectious diseases" (Curriculum Planning & Development Division, 2020, p. 24)
Pre-University (H1)	Biology	Impact of Climate Change (Extension Topic)	Students will explore how "way of life influences climate change" and how climate change links to the evolving "threat of infectious diseases" (Curriculum Planning & Development Division, 2020, p. 24)

Summarised from (Goh, Tan, Chang, & Ooi, 2009) and updated based on the syllabus documents in 2016 and 2020.

non-examinable at the Primary School Leaving Examinations (PSLE) and these subjects are considered to have low stakes compared to other subjects such as English Language, Mathematics, the sciences or the humanities, which are examinable subjects at the GCE "O"-levels. The implication of this is that the seriousness with which such issues are taught and learnt may be compromised.

While the layman's understanding of a subject discipline is a field of study, Shumway and Masser-Davidow (1991) were unable to obtain a satisfactory definition of discipline from their literature review and posit that "discipline of disciplinarity remains as yet undisciplined". Indeed, each discipline has a unique discourse on explaining phenomena and this perspective is influenced by how the discipline has developed historically. In Wilson and Wineburg's (1988) work on "Peering at History through different lenses: The role of disciplinary perspectives in the teaching of American history", the authors examined how social studies teachers' diverse disciplinary backgrounds influence the way these teachers thought "about history" and consequently effected the way they teach history. Indeed, learning about subject matter is not merely equivalent to gaining new knowledge about the subject but it also involves "examining previously held beliefs" (Wilson & Wineburg, 1988, p. 529) by teachers.

In particular, teachers from different disciplinary backgrounds have varying understanding of "the role of factual knowledge, the place of interpretation, the significance of chronology and continuity, and the meaning of causation" (Wilson & Wineburg, 1988, p. 527). The long description of the conception of disciplines is deliberate in that the topic of climate change draws on multiple disciplines. While, we have established earlier the key concept of uncertainty in the scientific, economic and political discourses of climate change, the issues can be further examined from geographical, historical, sociological and cultural perspectives. Indeed, a geographer will teach the topic of greenhouse warming very differently from a historian.

While I will not make any evaluation of the relative comparative advantages of each discipline on distinctive aspects of CCE, the case of Singapore can be used as an example for discussion in considering how CCE can be parked within a formal school-curricular subject.

CCE through school subjects

At the upper secondary level, the subject of geography is where CCE is naturally found as it aims to develop an informed concern in students about the quality of the environment, the use and management of resources and the importance of sustainable development at the local and global levels (Singapore School Examinations & Assessment Board, 2010). Indeed, the legitimacy of Geography as a subject that rises to the occasion in school curricula to become a vessel for the delivery of CCE is further strengthened by the inclusion of related environmental topics.

Singapore is a low-lying island with a coastline of 217 km in 2020 (Department of Statistics, 2022). This was a 111 km increase from massive land reclamation to the 106 km coastline in the 1970s (Chia, Khan, & Chou, 1988). Classified as a small island state by IPCC, coastal resources could be severely compromised by the imminent threat of human-induced global-warming-related climate change. A sea level of 0.2–0.86 metres is estimated to cost more than US$16 million for Singapore to protect its coast at the turn of the next millennium (Ng & Mendelsohn, 2005). It is unsurprising that a topic on Coasts is featured in both the past and current syllabuses. Students learn about how the coastal landforms are shaped by the physical processes, understand the impact of coastal erosion on human livelihood and discuss and evaluate coastal protection strategies. Interestingly, as the Singapore government responded to the concerns of sea level rise, the topic of coasts was re-introduced in 2007 after having been taken out of the syllabus in the 2001 syllabus. As an island state, most of its economic activities are located within two meters above mean sea level. With a gradual but imminent prospect of being inundated, Dutch consultants were engaged to combat rising sea levels. The government is "already in consultations with Delft in Holland to learn how [they] can build dikes" (Arnold, 2007). As Singapore's "state legitimacy is largely derived from the state's ability to develop the country" (Neo, 2007, p. 186), the environmental costs are a formidable challenge for the government. Perhaps this was the impetus that precipitated the reintroduction of the topic coupled with a more concerted effort at raising environmental awareness to educate the public.

Over the last 40 years, the topic of weather and climate has been taught at the upper secondary levels. Despite the rather monotonous tropical weather near the equator, making it challenging to engage student interest of the topic, the topic of weather and climate has stayed within school geography. However, there has been a rekindling of interest to teach climate change. The topic of global warming was only included in the lower secondary syllabus before 2012 but in the new syllabus for the upper secondary geography in 2013, the topic of climate change was included (Chang, 2012c) (Table 3.6).

In summarising the place of CCE within the formal curricula, Table 15 shows that there are gaps in the table arising out of the analysis of syllabus documents. These areas are normally taught by teachers through what they call "hidden curriculum". It would be prudent to analyse other sources of information to understand where CCE is taught and learnt in the formal curriculum. It is interesting to note that CCE cuts across a key topic in the upper secondary geography subject in the newly implemented curriculum as a unit on "variable weather and changing climate". The outcome can be attributed to the close collaboration between the National Climate Change Secretariat (NCCS) and the Curriculum Planning Design and Development unit at the education ministry to safeguard the topic's inclusion in formal education (Chang, 2012c).

Singapore is a signatory to the UNFCCC but only ratified the Kyoto Protocol in 2006. A revision to the Singapore Green Plan 2012 was adopted in 2006 with clear objectives of cutting carbon emissions. The inter-ministerial NCCS was formed under the Prime Minister's Office in 2010 for coordinating the climate change management strategies for Singapore as a whole. Although it is concerned with educating the public and raising the general level of awareness on the issue, the NCCS played an influential role in the

Table 3.6 Summary of the climate change learning outcomes in the formal curricula in Singapore

	Causes of climate change	*Impact of climate change*	*Managing climate change*
Knowledge • Recall	Sec 1,2 Soc Studies	Sec 1,2 Science	Sec 1,2 Science Sec 1,2 Geography
• Understand	Sec 3,4 Geography Pre-U Geography	Sec 3,4 Geography Pre-U Geography Pre-U Biology	Sec 2 Geography Sec 3,4 Soc Studies Sec 3,4 Geography Pre-U Geography
Skills • Analyse	Sec 3,4 Geography Pre-U Geography	Sec 3,4 Geography Pre-U Geography	Sec 3,4 Geography Pre-U Geography
• Evaluate		Sec 3,4 Geography	
Values			Pri 4,5,6 Civics Moral Education Pri 4,5,6 Science Sec 3,4 Soc Studies Sec 1,2 Science Sec 3,4 Geography

syllabus development of school geography (Chang, 2012c). Apart from the potential economic cost to the developmental state, Singapore also realises the possible impact on the physical and social environments.

In Singapore, geography is not a compulsory subject at the upper secondary level. It is compulsory for students to study a compulsory Combined Humanities subject, which comprises two papers. The compulsory paper is followed by a second elective humanities paper, where students can choose to study History, Geography or Literature. Students who choose a History or Literature Elective may also select to study a second humanities subject such as Geography. Most students study Social Studies or "Combined Humanities" than "pure" geography. School students come into contact with topics of sustainable development, including natural resources management and global warming through the social studies subject. Nonetheless, in the upper secondary SS syllabus, students are required to describe, explain and evaluate the strategies. It requires students to develop attitudes similar to those in the lower secondary SS syllabus. However, how these can be achieved is left to the teachers' discretion when enacting the curriculum.

In short, both the lower and upper secondary SS syllabi target knowledge, skills and values required for Education for Sustainable Development, while the primary school SS syllabus addresses the big ideas of human-environment interaction and environmental stewardship. CCE is only existent in the "pure" Geography subject and it includes the entire spectrum from learning about the causes, impact and management:

- How has global climate changed?
- What causes global climate change?
- What is the greenhouse effect and how does it work?
- What is the enhanced greenhouse effect?
- How do human activities lead to the enhanced GHE?
- How does climate change affect people?
- How do countries respond to climate change?

(Singapore School Examinations & Assessment Board, 2012a)

While the first topic on how the global climate has changed draws on students' ability to recall information about how global temperatures have risen over different time scales, it sets the stage for the discussion on how it will impact human activity as the changes in temperature over the last 160 years have been rapid and unprecedented (Figure 3.2).

The next four topics on explaining the causes of climate change, explaining the mechanism of greenhouse warming and attributing it to an enhanced greenhouse effect and evaluating how humans have contributed to the increase in greenhouse gases are all consistent with the development of scientific and political discourses as outlined in Chapter 1. Increasingly, public discourse on CCE has been incorporated into the curriculum through the inclusion of these topics. In other words, the curriculum intends to level students up to the scientific and political discourse, providing them with the

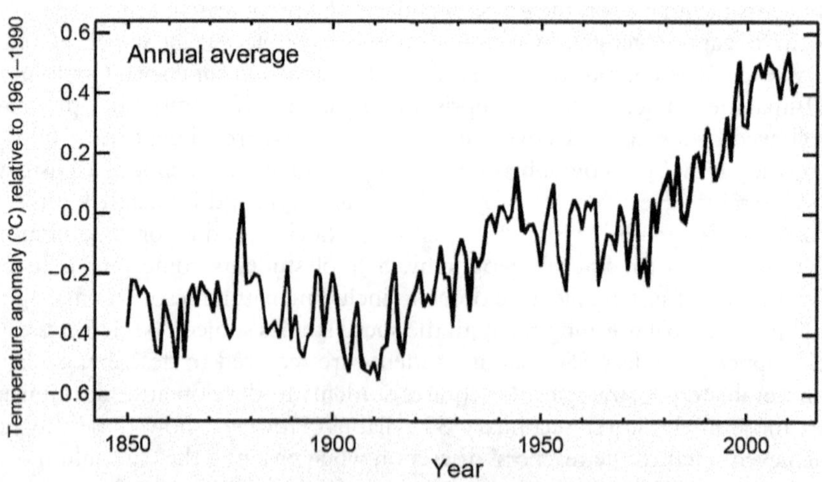

Figure 3.2 Global mean land–ocean temperature index, 1850–2014.

Source: Adapted from National Aeronautics & Space Administration (2012) & Stocker et al. (2014)

knowledge and information required for critical inquiry, so that they can be active citizens.

On the topic of how climate change affects people, students are required to recall and understand the environmental impact of climate change. Moreover, they will need to analyse and evaluate how these environmental changes, such as increasing average temperatures, will affect the social and economic activities. For instance, apart from learning about the potential inundation in low-lying areas due to sea level rise, students have to evaluate how this will affect the livelihood of fishermen.

The curriculum also intends for students to learn about the international political discourse on climate change. To some extent, it levels students up to the debates that surround the uneven adoption of mitigation strategies across different regions. In addition to recalling and understanding the management strategies, students have to analyse, evaluate and even form an opinion about how efficacious the international strategies have been in mitigating climate change.

In Geography, at the pre-university level, CCE is only found at the Higher 2 (H2) level (H2 level is comparable to the difficulty and rigour of the previous GCE "A"-level). The knowledge, skills and values for this unit target the core concepts in CCE encouraging students to go beyond awareness and to take action. Conceptually, there is no discernible continuation of climate change issues from lower secondary to upper secondary geography syllabi in terms of skills and values, although there is a distinct section on environmental issues in Social Studies, which is compulsory for upper secondary levels.

Holistic education versus academic agenda

Singapore's Desired Outcomes of Education (DOE) refer to attributes that we would like to observe in children "by the completion of his formal education" (Ministry of Education, Singapore, 2009). A student who has attained the DOE would have

> a good sense of self-awareness, a sound moral compass, and the necessary skills and knowledge to take on challenges of the future. He is responsible to his family, community and nation. He appreciates the beauty of the world around him, possesses a healthy mind and body, and has a zest for life.
>
> (Ministry of Education, Singapore, 2009)

A student from the Singapore system should also have an inquiring mind, critically evaluating the information that he or she is presented with. To this end, the Singapore stage seems to be set for CCE, in that it supports the care and criticality that both ESD and EE espouse. Further, the goal of CCE is to *help learners develop knowledge, skills, values and action to engage and learn about the causes, impact and management of climate change.*

Whether it is ESD or CCE, both of these are assumed to be addressed together in the environmental education component of the school syllabus. In fact, environmental education with a slant towards sustainable development was the main focus provided within the much earlier syllabi and only recently has the issue of climate change come to the forefront and made explicit in the revision of the curriculum. Nevertheless, in the minds of the curriculum planners, and understandably so, both issues are intertwined and not mutually exclusive. While CCE exists in many pockets within the school curriculum, in tandem with a slightly more prominent presence of ESD, the subject curricula are articulated through national-level syllabi, which are then interpreted by teachers at the school level. The interpreted curriculum is the one that is being implemented within the classroom. While state-driven education policies are interpreted within the social context of a nation where people are considered generally law-abiding, the variation in interpretation will invariably result in differences in the way CCE is carried out.

From the description of Environmental Education in formal curricula, it can be seen that the concerns of the Ministry of the Environment and Water Resources related to pollution, conservation of resources, energy and the environment and the effects of climate change are being addressed through the formal education of the students.

However, the focus is mainly on the cognitive level in the sciences and the social sciences and may remain as just knowledge to acquire to pass examinations. Students are provided opportunities to act on their school learning about sustainable development in community projects where they are actively involved in "doing something" for the environment and feel that they can make a difference to the environment. Community-based environmental programs complement and supplement formal Environmental Education, allowing students to be personally involved and in making a tangible difference – education *for* the environment.

Issues of climate change have become more prominent in Singapore over the last decade because the leadership in government is discussing them in the public arena, and the momentum in focusing on and addressing them will only accelerate in the future. This recognition of climate change and environmental impact of various kinds due to human actions will influence schools to increase their students' focus on them. Already there are some promising signs that this is happening. There are now more frequent and positive reports noting that primary school children are examining environmental issues more frequently in topics such as climate change (Gilani Research Foundation, 2009). However, the question remains as to how efficacious is such as two-part treatment of a child's environmental education.

In summary, CCE is a crucial component of any effective climate change mitigation approach. Education is seen as a tool that may help people and groups make informed choices about how to respond to the challenge posed by climate change and increase their adaptability abilities in preparation for the effects of climate change as they become more apparent. It is essential to recognise the strategic role that education plays in the fight against climate change, not only in the area of adaptation but also in the area of mitigation. In this chapter, we discussed the ways in which Singapore's existing educational system addresses concerns relating to climate change as well as to related topics. Using the example of Singapore was intentionally done so that it could be compared to studies done in other parts of the world. In Singapore, CCE is often delivered via the school subject of geography, typically at the upper secondary level. The purpose of this subject is to cultivate in students an informed concern about the quality of the environment, the use and management of resources and the significance of sustainable development at both the local and global levels.

Studies have shown that there is often gaps between knowledge, attitude and action (Andersson & Wallin, 2000; Lee, Lester, Ma, Lambert, & Jean-Baptiste, 2007). Despite these gaps in learning for CCE, and before we can advance a framework to implement CCE, it would be important to understand what our learners know or do not know. Only then will we be able to design curriculum and instruction that will engage the student meaningfully in learning about climate change, bridging the gap between knowledge and action.

4 What do students know about climate change?

Climate change education (CCE) as a mitigation strategy for climate change is contingent on the effectiveness of the curriculum. In turn, the curriculum can only be designed and implemented successfully if we understand where students are at, in terms of their knowledge and disposition about the issue. Indeed, research has shown that students are more engaged and consequently more successful if they are taught in ways that correspond to their readiness levels (Vygotsky, 1986). It is through identifying what they know and what they do not, that CCE can be more effectively implemented. Inherent in this position is the assumption that there are gaps in knowledge of learners and that we need to identify misconceptions in knowledge transmission and acquisition. Thus, this chapter will examine the literature on how much students know about climate change and report on the results of a study conducted on the level of understanding of students about climate change in Singapore.

What do students know about climate change?

In considering the framework developed in this book on the goals of CCE, the evaluation of what students know about climate change has to consider how *learners develop knowledge, skills, values and action to engage and learn about the causes, impact and management of climate change*. To this end, the following review will not just focus on the knowledge and understanding of climate change but also examine the types of attitudes and the kind of action that they take.

A survey of the literature shows that students' knowledge of climate change is predominantly erroneous and lacking in depth (Lee, Lester, Ma, Lambert, & Jean-Baptiste, 2007). Not dissimilar to the public confusion over the issue, students essentially perceive the hole in the ozone layer as the main cause of global warming (Andersson & Wallin, 2000). Such a misconception is persistent across age groups and year levels in different research locations, such as in the United States (Gowda, Fox, & Magelky, 1997; Lee, Lester, Ma, Lambert, & Jean-Baptiste, 2007; Cordero, Todd, & Abellera, 2008), Norway (Hansen, 2010), the United Kingdom (Boyes & Stanisstreet, 1993; Francis, Boyes, Qualter, & Stanisstreet, 1993; Jeffries, Stanisstreet, & Boyes, 2001;

DOI: 10.4324/9781003093800-4

Daniel, Stanisstret, & Boyes, 2007), Greece (Koulaidis & Christidou, 1999; Padpadimitriou, 2004), Sweden (Andersson & Wallin, 2000), Turkey (Kilinc, Stanisstreet, & Boyes, 2008) and Singapore (Chang, 2012b; Chang & Pascua, 2016; Chang, Pascua, & Ess, 2018). In fact, their confusion extends to their basic understanding of concepts such as the weather and the climate (Gowda, Fox, & Magelky, 1997; Padpadimitriou, 2004).

In terms of climate change impact, there is a tendency for students to inflate the estimated temperature changes and conclude that the effects of climate change in the short term, in rising temperatures, are easily detectable (Gowda, Fox, & Magelky, 1997). They are predisposed to interpret environmental issues such as lead pollution, radioactive contamination and acid precipitation as calamities that could be related to and explained by climate change science (Boyes & Stanisstreet, 1993; Francis, Boyes, Qualter, & Stanisstreet, 1993; Gowda, Fox, & Magelky, 1997; Padpadimitriou, 2004). Students commonly have the impression that all pro-environment strategies are appropriate to solve the challenge of climate change (Boyes & Stanisstreet, 1993; Fortner, 2001; Daniel, Stanisstret, & Boyes, 2007; Punter, Ochando-Pardo, & Gracia, 2011). Further, some students have the belief that the greenhouse effect (GHE) causes skin cancer (Jeffries, Stanisstreet, & Boyes, 2001).

Further, old beliefs could get deeply ingrained over time (Lee, Lester, Ma, Lambert, & Jean-Baptiste, 2007). Following the interventionist pre-/post-assessment design, it is discovered that in a sample of fifth-grade students across the United States, learners continued to have difficulty differentiating between the GHE and the enhanced GHE, as well as ozone layer depletion as a different concept, even after discussion of such topics for post-assessment.

Formal education seems to have worked in some cases where students have demonstrated that the main messages of climate change and its impacts have been clearly delivered to them after the lesson. Indeed, positive attitudes towards climate change and their knowledge of climate change and attitudes are interlinked (Grant & Featherstone, 2009). It appears that families' and schools' support is the key to helping students feel "empowered to make climate-friendly changes" in their behaviours (Grant & Featherstone, 2009).

Various researchers note the persistence of long-held beliefs and varied interpretations about the science of climate change. Whether interventionist approaches are able to correct these misconceptions concerns scholars. Cordero, Todd, and Abellera's (2008) evaluation of a 15-week course on weather and climate among university students in California indicates that there was a notable lack of deeper learning involved and the course failed to change the misconceptions of students. It is believed that these misconceptions could be traced to how the topic of climate change was covered in lower year classes and further exacerbated by media reports. The authors then posit that there should be revision of curriculum design and specific alterations to be introduced to correct erroneous ideas that are passed onto students.

In an investigation of how a change in the curriculum of high schools in the United States supports students' development of "critical science agency", McNeill and Vaughn (2012) found that at the end of the programme, the students displayed substantial improvement in understanding of climate change as well as behavioural and attitude change (McNeill & Vaughn, 2012). Conceptual understanding of climate change is a prerequisite to understanding the impact of one's personal actions. It is then recommended that changes in the curriculum must work on supporting students' learning by addressing misconceptions and motivating students for pro-environment action.

Mason and Santi (1998), in their investigation of fifth graders' conceptions of the GHE and global warming, suggest that collaborative learning has the potential of building up new knowledge, a viable pathway to revise one's personal conceptions and beliefs (Mason and Santi, 1998). As the model encourages meaningful negotiation and sharing, there is collaborative knowledge construction. As the group engages in argumentative dynamics, children become more explicit in their reasoning and presuppositions. This dynamic promotes metaconceptual awareness, aids in deconstructing latent knowledge and in the reconstruction of new and improved understanding. Collaborative learning, as the authors maintain, distributes the cognitive burden of thinking and opens opportunities to dissect and critique existing beliefs and knowledge.

How such unconventional ideas are generated and become commonly held beliefs concern some researchers. In studying how students make sense of data about climate change, Svihla and Linn (2012) explicate that ideas are formed from everyday experiences, observations and the popular media. Students connect concepts learnt from school with other positions gathered from alternative sources to support an assertion even if these new ideations are not directly related. These assertions could be grounded on everyday experiences mixed with some science concepts. Sternäng and Lundholm's (2011) interviews with 14-year-old students confirm that students conceptualise the solutions to mitigating climate change in relation to two different stances – either themselves or from another person's point of view. Thus, views on the matter are rather fluid; views change accordingly in relation to a person's social context.

Nonetheless, a research by Braten, Stromso, and Salmeron (2011) showed that students perceive data from textbooks and official documents to be trustworthy more than information from newspapers and other sources. Results also showed that readers with low topical knowledge tend to trust less reliable sources and could not differentiate the relevance of the criteria when judging the trustworthiness of sources (Braten, Stromso, & Salmeron, 2011).

The prevalence of misconceptions, confusions and alternative explanations in the understanding of the global climate change phenomenon merit insights and suggestions by various researchers (Chang & Pascua, 2016). Bangay and Blum (2010) maintained that education is an untapped resource to contribute to adaptation and mitigation measures in penetrating mainstream development thinking. The trend of policy discussions has been, by far, focused on technical solutions without seriously engaging with the content of education. Svihla and Linn (2012) posit that in-school instruction is vital in guiding students to generate their own ideas. It is a venue for them to get feedback about their views and experiments. Finally, students are able to comprehend complex systems, the global climate system, for example, when the curriculum is carefully designed. It underscores the importance of using evidence from student work to guide iterative refinement of instruction.

The need for effective teaching strategies in bringing the issue of climate change into the classroom is vital (Boyes & Stanisstreet, 1993). It has been the trend that climate changes, particularly global warming and the GHE, are integrated into existing school subjects. In Greece, for instance, the GHE is not explicitly included in the curriculum. Related concepts and ideas are integrated into the textbooks in several cases and in different contexts (Koulaidis & Christidou, 1999). This patchy coverage thus leads to a "lack of rich conceptualization" in the understanding of climate change (Shepardson, Niyogi, Choi, & Charusombat, 2009).

Case study: Singapore

As the studies cited are outside of the Asian context, there was a need to ascertain what Singapore students know about climate change given the dearth of empirical studies on CCE for the island state. In order to have a meaningful discussion of the curriculum for CCE, its implementation and even offer suggestions about instructional strategies, students' current state of understanding of the topic of climate change has to be determined. Two studies by the author (Chang, 2012b; Chang & Pascua, 2016) will provide a glimpse into what students know or do not know about climate change.

Chang (2012b) employed a mixed methods approach involving various iterations with an aggregation of qualitative and quantitative data gathering techniques to assess students' knowledge about climate change while Chang and Pascua (2016) attempted to determine whether classroom instruction can successfully correct learners' misconceptions about climate change. Chang's (2012b) research was a baseline study which made use of the exploratory and descriptive approaches to inquiry.

Baseline study of students' knowledge about climate change

A performance task and survey were administered to determine what students know about climate change. The performance task was composed of questions that draw out students' understanding of basic climate change concepts. The survey used Likert-type questions meant to assess learners' climate change awareness, their attitude and behaviour towards CCE and their sources of information on climate change.

Eleven secondary schools ranging from public, private all-girls and all-boys schools participated in the research. Given the challenge of gaining access to schools, random sampling was impractical and purposive sampling was employed instead, to ensure a good spread of high- and average-performing schools in the frame, as well as a balance of resource-rich schools with that of public schools. While it was reiterated to the teachers that sampling should be representative of the clusters, which means there is equal representation among the lower and higher ability groups, they were given free rein to handpick which cohorts and students they included in the list. A total of 397 students belonging to the Secondary 3 cohort participated in this exercise.

Most (87.7%) of the respondents belong to the higher ability stream called the Express stream and the rest belong to the Normal Academic/Technical streams. Singapore attends at least ten years of compulsory general education. After six years of compulsory primary education, a student will have to be enrolled in a four- or five-year secondary education programme. Express stream students take the GCE O Level examination at the end of four years while Normal stream students do so only in their fifth year. There was an almost equal number of boys and girls with a slight majority in the latter. The sample was composed mostly of Singaporean citizens (84.6%) with some permanent residents (7.6%) and international students (6.5%). The ethnicity distribution reflected the normal population spread in Singapore, composing mostly of students of Chinese descent (75.8%) followed by Malays (11.8%) and Indians (6.3%). The relative proportions of nationality and ethnicity are illustrated in Figures 4.1 and 4.2.

Twenty-seven students who took the survey were then invited to take part in either one-on-one or paired interviews. Similarly, Focus Group Discussions (FGDs) with teachers were also organised to discover how CCE is perceived and taught by educators. A total of 40 teachers participated in one of seven FGDs and two interviews. The aim of these conversations was to

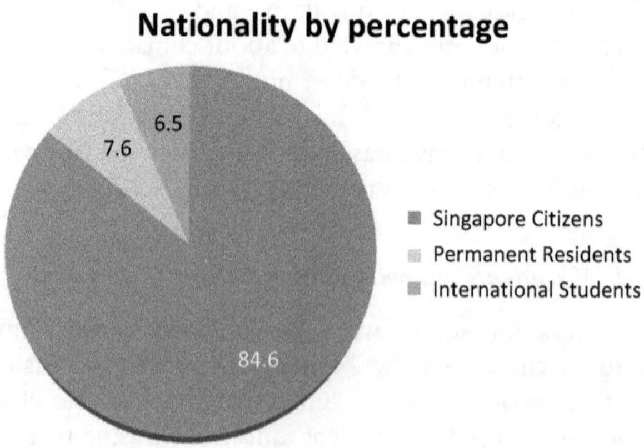

Figure 4.1 Distribution of survey participants by nationality.

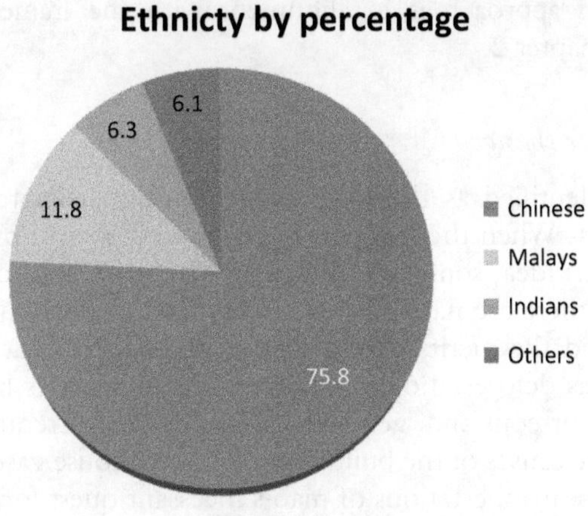

Figure 4.2 Distribution of survey participants by ethnicity.

corroborate data from the survey and to broaden the discourse through contextualised perspectives.

A research team composed of the principal investigator, two research assistants along with six student assistants systematically coded answers to the open-ended questions in the performance task. The student assistants are university students who took a course on climate change taught by the principal investigator. The coding assignment was structured for three (3) persons to work for one item: two (2) main coders and one (1) arbitrator. Prior to the coding process, the team discussed at length the procedure, codes, concepts and criteria for coding. Two persons proceeded to code raw data independently. Inter-rater reliability (IRR) analysis was then employed using the Kappa statistic to confirm consistency of rating and to check for researcher's bias. In events of non-agreement on an item, the arbitrator stepped in to facilitate discussion and resolution of the issue.

The learning of climate change

Learners' reading of climate change reflects what observers such as Shepardson, Niyogi, Choi, and Charusombat (2009) call "lack of rich conceptualization". Their understanding is defined by rote memorisation and is limited for the purposes of taking an examination. As one informant stated,

> We do not learn how to interpret data so, so in depth. It's on the surface only, as in like, the basic things of how climate change affects us. But we don't really, really go and research on why it happens.

The discussion that follows shows the gist of data gathered on students' understanding, behaviour and attitude towards climate change and the learning of the causes, effects and management of climate change. This purposeful dual dimension approach is in alignment with the framework for CCE described in Chapter 2.

Causes of climate change

Humans are identified as the main contributor to climate change either solely or in part. When the students were asked if they know of alternative thoughts on this idea, some acknowledged that they have come across the argument that there are natural causes of climate change but have not seriously considered the merits of the case. It is believed that anthropogenic activities such as deforestation, emission of harmful gases by factories, the use of cars and air-cons and, generally, all activities that require the burning of fossil fuels are causes of the building up of greenhouse gases in the atmosphere. These are manifestations of man's incessant quest for modernity and for some students, a blatant sign of the human race's greed. Students explained that as people become more and more dependent on technology, it is inevitable that more electricity is consumed to run machines and power cities, further increasing the excessive consumption of fossil fuels. This then feeds into the cycle of releasing more gases that harm the Earth finally resulting in changes to the climate system such as global warming.

The GHE or the enhancement of it is listed as a key element for drastic changes in the climate. Associated concepts to this are the trapping of heat, the increase in volume of greenhouse gases and the excess carbon dioxide being released into the atmosphere. One of the items in the performance task given to the students required them to provide a definition and explication of what the enhanced GHE is. The question posed was

> What is the enhanced greenhouse effect? Please describe briefly the mechanism through which the enhanced greenhouse effect takes place by specific reference to the individual compounds that are responsible for it.

The expected answer should encompass, but is not limited to the material of the following:

> The enhanced greenhouse effect refers to an enhancement of the natural greenhouse effect. Energy from the sun is first absorbed by the Earth and the Earth will eventually lose this energy gained. The Earth loses this energy through long wavelength radiation (infrared). Naturally, the gases in the atmosphere such as Carbon Dioxide and Water Vapour will block some of the long wavelength radiation, hence trapping the out-going radiation and warming the Earth. This is called the natural greenhouse effect. The enhanced greenhouse effect refers to an enhancement in terms of the amount of and types of greenhouse gases (e.g. Nitrous Oxide). This increase in amount and type of greenhouse gases will exacerbate the warming.

Table 4.1 Categories of responses for students' understanding of the enhanced greenhouse effect

Categories	*Expected response*
No understanding	No evidence of any knowledge & understanding of what the enhanced greenhouse effect is
Wrong understanding	Knowledge is evident but misconceptions exist or some facts are erroneous. Mentions the trapping of heat but is unable to explain the mechanism for heat trapping OR no distinction from the natural greenhouse effect
Limited understanding	Knowledge is evident and most information is accurate. Able to distinguish between enhanced and natural GHE. No explanation for the mechanism
Accurate understanding	Understanding of what the enhanced greenhouse effect is clear, with explicit mention of the mechanism for how heat is being trapped and how it differs from the natural greenhouse effect

Results showed that more than half of the sample (54.4%) have ***no understanding*** of the concept. In context, "no understanding" was assigned to answers that are irrelevant to the question asked (e.g., "An updated house that is green in colour"), drawings that suggest no connection to the subject and one-liners that do not answer the question (e.g., air pollution, global warming). Included in this category were direct answers that confirm no knowledge about the matter (e.g., I don't know; I don't understand). In addition, a significant number (30.5%) provided erroneous answers, thus, with ***wrong understanding***. Though some concepts were correct (e.g., trapping of heat), misconceptions were also prevalent such as the trapping of sunlight/inability of sunlight to escape into space and ozone depletion causing the enhancement of the GHE. Explanations provided were likewise devoid of differentiation between the natural against the enhanced GHE. Those with ***limited understanding*** (6.3%) were able to point out the differences between the two but the explanations provided were usually incomplete. For instance, they failed to incorporate and discuss elements such as infrared radiation, water vapour and carbon dioxide, or they merely mentioned one or two of these factors in passing. Table 4.1 shows a summary of the categories of responses from the students.

Similarly, when the students were tested on whether they are aware of and are able to argue that climate change may be due to either natural or anthropogenic causes or both, it was found that they do not have the facility to reason beyond common perceptions. The following performance task was posed to the students:

> The news report below presents arguments against common understanding on global warming. Do you think that global warming is due to human activity? Why/Why not?
>
> Newspaper article: The arguments made by climate change sceptics

> Instruments show there has been some warming of the Earth's surface since 1979, but most data comes from surface weather stations, which are urban centres. When these stations observe a temperature rise, they are simply measuring the "urban heat island effect". In addition, coverage is patchy, with some regions of the world almost devoid of instruments. Data going back further than a century or two is derived from "proxy" indicators such as tree-rings, which are subject to large errors.
>
> The argument that CO2 leads to global warming is also challenged through ice-core records. Dating back nearly one million years ice-cores show a pattern of temperature and CO2 rise at roughly 100,000-year intervals. But the CO2 rise has always come after the temperature rise, not before, presumably as warmer temperatures have liberated the gas from oceans. Further, Earth history shows climate has regularly responded to cyclical changes in the Sun's energy output. Any warming we see can be attributed mainly to variations in the Sun's magnetic field and solar wind.
>
> (Adapted from BBC Online news, 13 Dec 2009)

More than half of the respondents (52.6%) took a stand on the issue, mostly saying that human activities cause global warming. They provided brief explanations listing down human activities known to damage the environment such as deforestation and the use of CFC-containing products. Those who argued that it is due to natural cause link back their position to the article provided, particularly quoting the "Sun's energy output" as a reason. Some stated that it is both a natural and human-induced process but the explanations provided only elaborated on one side of the debate. A good number (36.5%) took a stand on the matter but provided no explanations at all. They merely stated "Yes", "No", "Maybe" or "Partially". Some acknowledged both sides of the debate but failed to elaborate further. In smaller percentages, some students acknowledged and argued for both sides of the debate (4.5%) and even evaluated the final stand taken (2.0%).

Those categorised under "No understanding" (1.3%) were answers of "I don't know" or when the students merely described how global warming leads to an increase in temperature without stating whether this is due to human activities or natural causes. This indicates that students remain uncritical in determining whether climate change is caused by human activity. Table 4.2 provides a summary of the responses for this item.

Effects of climate change

Temperature increase is unanimously perceived as the most likely result of climate change. With hotter Earth temperature, it is perceived that the ice caps in the Polar Regions will melt, leading to a surplus of water that will eventually result in sea level rise. This spells catastrophes like floods in some parts of the Earth and droughts in other areas. Such climatic changes

Table 4.2 Responses to news article to determine if climate change is natural

	Frequency	*Percentage (%)*
No understanding	5	1.3
Takes a stand, with no explanation	145	36.5
Takes a stand, with an explanation	209	52.6
Takes a stand, argues for both sides of the debate	18	4.5
Takes a stand, argues for both sides of the debate, evaluates the final stand	8	2.0
No answer	12	3.0
Total	397	100.0

will bring about loss of life and damage to infrastructure and properties. Low-lying coasts and small island nations such as Singapore are believed to be inundated as a result.

The performance task that asked the student to mark all events in a list that they think will occur as a result of climate change, corroborated the main ideas gathered in the student interviews. Almost all students are absolutely convinced that there will be a rise in global temperature (95.5%) and that many countries' climates will change (91.9%). They also believe that there will be more floods (87.4%) and the polar ice caps will melt (98.0%). However, they were divided over items such as the formation of more deserts and the possibility of crops suffering more plagues. The consensus leaned more on the perception that these are plausible results of climate change (more than 50%) by a margin over those who think otherwise.

Most notably, students were not convinced that there would be fewer typhoons as a result of climate change (92.2% for False) while three in four think the event will not result in river contamination. Items not related to climate change were also included in the questionnaire to test whether students were able to differentiate between purely natural catastrophes and the ones that may be caused or aggravated by climate change. Interestingly, roughly one in three students think volcanic eruptions and earthquakes may be triggered by climate change, which indicates an area where student misconception exists (Table 4.3).

They were then asked to pick one of the choices above and discuss how changes in the climate contribute to the likelihood of such an event happening. More than half of the respondents (54.9%) had *accurate understanding* of how climate change results in certain events. For example, they were able to discuss the concepts of temperature increase as a result of the enhanced GHE that then leads to melting of the ice caps, or increased evaporation leading to heavier rainfall and thereby floods. Those with *no understanding* (23.4%) gave irrelevant answers such as "Increasing sea levels will flood factories thereby contaminating rivers with the waste from the floods". Others expressed a complete lack of knowledge on the matter (I don't know/I don't understand). Consistent with the earlier pattern seen in the enhanced

Table 4.3 Effects of climate change

Consequences	*True*	*False*
Earth's temperature will increase	379 (95.5%)	17 (4.3%)
River contamination will increase	98 (24.7%)	298 (75.1%)
Crops will suffer more plagues	215 (54.2%)	181 (45.6%)
There will be fewer typhoons	30 (7.6%)	366 (92.2%)
Many countries' climates will change	365 (91.9%)	31 (7.8%)
There will be more floods	347 (87.4%)	49 (12.3%)
The polar ice caps will melt	389 (98.0%)	7 (1.8%)
There will be more deserts	219 (55.2%)	177 (44.6%)
There will be more volcanic eruptions	124 (31.2%)	272 (68.5%)
There will be more earthquakes	142 (35.8%)	254 (64.0%)

GHE question, answers pertaining to the hole in the ozone layer significantly contributing to global warming persisted, thus *wrong understanding* (15.1%). Some have *limited understanding* (3.8%) with incomplete answers and insufficient explanations such as "Hot countries get hotter, cold countries get colder" or "Temperature change hence ice caps melt" without mentioning whether there is temperature increase or decrease involved.

While erratic weather patterns such as storms and hurricanes were indicated as key effects of climate change, students do not know the processes involved in such events. For instance, they (38.5%) have *limited understanding* of how hurricanes are formed. Some were able to identify that the increase in temperature as a result of global warming results in an increase in evaporation but failed to mention that it is the latent heat generated that fuels the formation of hurricanes. There were those who indicated briefly how the increase in temperature may cause hurricanes but provided no explanation for it. There were also several respondents who believe that hurricanes are due to natural causes. Additionally, 31.7% have *no understanding* – the respondents simply wrote "Yes", "No", "I don't know" or mentioned climate change as a major factor but failed to elaborate any further. Those with *wrong understanding* (18.6%) mostly failed to recognise that several factors jointly contribute to the formation of hurricanes. Some are even under the impression that hurricanes are caused solely by storms and strong winds.

In a question to test for the students' explication of the impacts of a hurricane that is possibly caused by global warming, a figure showing the short-term impacts of Hurricane Katrina on the Gulf Coast of the USA was provided and students were asked "Do you think this is due to global warming? How can it impact humans?" A large proportion of students (36.0%) opted not to give an answer. This is despite the fact that the diagram offered information to base their opinions on. Another 13.1% wrote, "I don't know" or stated devastations caused by storms or floods in a general manner. Those who did, mostly described impact (40.6%) with generic answers such as deaths, damage to infrastructure and the loss of homes. Others attempted to go beyond the description of impact to explain further the effects (5.8%). Such include elaboration on how damage to infrastructure and the loss of homes can lead to economic problems. Others referred to the diagram mentioning, "ports and harbours were wrecked" or "at least 50 people were killed".

Rather than testing students' prior knowledge of the impact of climate change on the environment and society, the item was designed to uncover if students were able to use the information provided to analyse and deduce the impact of climate change. In other words, the level of cognitive engagement was tested in this case to see if students could think critically. To be fair, these 15-year-old Secondary 3 students had little opportunity to learn it from the formal curriculum in their lower grade levels as the topic of the impact of climate change was not in their curriculum.

Climate change management strategies

The 3Rs (Reduce, Reuse, Recycle) figure as the top-of-mind mitigation strategy known to students. In the performance task, for instance, they were asked to give suggestions on how humans could counter the effects of a changing climate. The respondents stated one or more strategies, the most common of which is the 3Rs and saving electricity. Examples presented were the act of switching from air-conditioning systems to fans, the use of public transportation in place of private cars, campaigning to educate the public and participation in global events such as the Earth Hour. Some 12.3% went as far as to explain how useful these strategies are but failed to mention their limitations. A significant number (20.4%), however, were unable to identify any response strategy (Table 4.4).

Students mainly associate the 3Rs with reusing plastic bags from supermarkets. It is observed by the researcher in the interviews that the participants tend to give suggestions that do not directly involve their active participation. By and large, students would take the bus or the train, as they do not drive cars just yet. When asked about the reason why they turn off the lights, a good number gave answers that pertain to habitual acts, in that they have been taught from when they were young to do so, they were nagged by a parent, or for pragmatic reasons such as saving electricity. Some noteworthy answers include a suggestion of developing alternative energy sources to lessen Singapore's dependence on fossil fuels. Others suggest intensified public education, the building of levees and dikes and advocacy campaigns such as the Earth Hour to be more pervasive and frequent.

One item in the performance task aimed to draw out students' perceptions of the effects of engaging in common activities that require the burning of fossil fuel. A list of activities was provided and they were instructed to mark all that they deem to satisfy the criteria given. Table 4.5 provides the list as well as the response rates from the students.

Table 4.4 Evaluation of climate change response strategies

Responses	*Frequency*	*Percentage (%)*
Unable to identify any response strategies; no answer	81	20.4
States one or more strategies; no specific examples given	63	15.9
States one or more strategies with specific examples given	156	39.3
States and describes response strategies but with no specific examples given	16	4.0
States and describes response strategies with examples; no explanation or evaluation of the usefulness of stated strategies	49	12.3
States, describes and explains response strategies with examples; no evaluation of the usefulness of stated strategies	26	6.5
States, describes and explains response strategies with examples; evaluates the usefulness of stated strategies	6	1.5
Total	397	100.0

Table 4.5 List of activities that exacerbate climate change and students' responses to them

Activities	*True*	*False*
Play football	15 (3.8%)	380 (95.7%)
Ride a bicycle	22 (5.5%)	473 (94.0%)
Read in natural light	16 (4.0%)	379 (95.5%)
Not recycle rubbish	296 (74.6%)	99 (24.9%)
Play video games	164 (41.3%)	231 (58.2%)
Ride a motorcycle	323 (81.4%)	72 (18.1%)
Use air conditioner	347 (87.4%)	48 (12.1%)
Have a hot shower	158 (39.8%)	237 (59.7%)
Download files from the Internet	95 (23.9%)	300 (75.6%)
Turn heater on	223 (56.2%)	172 (43.3%)
Listen to mp3	99 (24.9%)	296 (74.6%)
Go to school by car	367 (92.4%)	28 (7.1%)
Use a hairdryer	211 (53.1%)	184 (46.3%)
Charge up a mobile phone	212 (53.4%)	183 (46.1%)

These activities were then grouped into three segments:

- The use of petrol-powered vehicles, which involves the direct burning of fossil fuels, such as going to school by car or riding a motorcycle
- The use of electricity-powered devices such as charging mobile phones, using hairdryers and playing video games.
- Engaging in activities that do not require the burning of fossil fuels, such as reading in natural sunlight, riding a bicycle or playing football

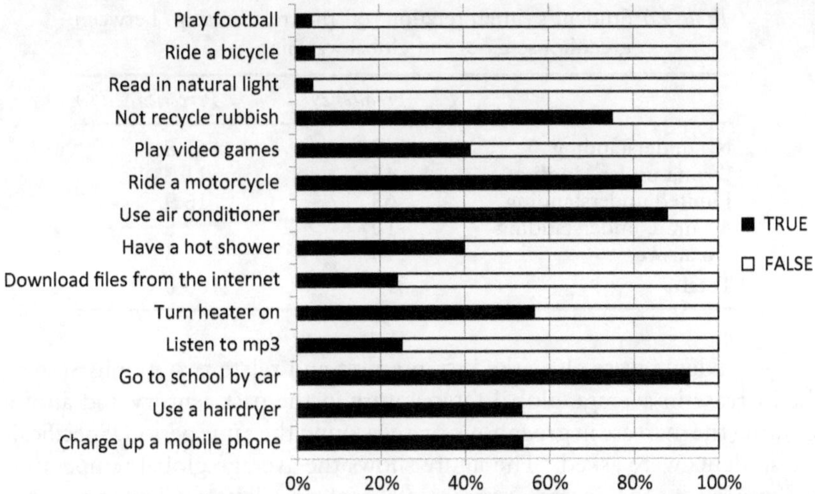

Figure 4.3 List of activities that exacerbate climate change and students' responses to them.

A great majority of students rightly distinguished that petrol-powered vehicles such as cars (92.4%) and motorcycles (81.4%) are anthropogenic contributors to climate change. Among the common household appliances listed, the air-con was singularly identified to hasten climate change (87.4%). Answers were mixed as to whether activities such as charging a mobile phone, using a hairdryer or a heater and playing video games could do much harm. Interestingly, three in four students believe that downloading files from the Internet could not possibly contribute to climate change. About one out of two could not see the impact of taking hot showers as well. Nonetheless, three out of four are convinced that not recycling rubbish may likely hurt the Earth. Almost all of the students identified the three activities pertaining to energy-efficient actions as non-contributory to the development of climate change (Figure 4.3).

In terms of their explications of how a common activity contributes to enhancing the likelihood of climate change to happen, more than half (52.6%) of the respondents were coded as having a *wrong understanding*. Though they supplied answers related to the question, misconceptions abounded in responses stating that cars emit toxic gases without relating such answers to the burning of fossil fuels. An activity such as air-con use was mainly understood to contribute to climate change due to the formation of chlorofluorocarbons that deplete the ozone layer; 25.7% of the respondents gave irrelevant answers, thus meriting a *no understanding* score. These answers are "I don't know/I don't understand", among others. Some (16.6%) have *limited understanding* with their insufficient responses. These included answers that do not fully illustrate how the burning of fossil fuels leads to the enhancement of the GHE and those that elaborated on the latter but failed to mention the former.

Table 4.6 Student's interpretation of the relationship between greenhouse gases and global warming

	Frequency	*Percentage (%)*
No understanding	269	67.8
Wrong understanding	45	11.3
Limited understanding	63	15.9
Accurate understanding	10	2.5
No answer	10	2.5
Total	397	100.0

The students were also asked to interpret and relate two graphs showing the increase in average global temperature in the past century and another showing the increase in greenhouse gases around the same period. Specifically, the students were asked "The figure shows the average global temperatures as reported by the Intergovernmental Panel on Climate Change report in 2007. What are the possible causes of this increase in average global temperatures?" (Chang, 2014, p. 60) (Table 4.6).

The majority of the respondents (67.8%) presented irrelevant answers – merely stating possible contributors to global warming such as cars, industries and carbon dioxide without drawing relations between and among these elements. Others opted to answer, "I don't know/I don't understand". These students were judged to have *no understanding* of interpreting the graph meaningfully. Some (15.9%) attempted to draw a relation between the increase in temperature and the increase in greenhouse gases, but the answers provided were incomplete; thus this group has *limited understanding*. The students also failed to integrate their analysis of the timeline into their arguments. About 11.3% have *wrong understanding* with misconceptions mainly traced to their erroneous knowledge that the depletion of the ozone layer significantly contributes to trapping of sunlight and, therefore, more heat within the Earth.

Students' range of cognitive and affective engagement in climate change

This book started by asking what we mean when we say someone has learnt something about climate change. The conceptual framework described in Chapter 2 has provided the need to examine the cognitive engagement of students beyond simple factual memorisation of climate change information. Indeed, the findings from students' knowledge in the domains of causes, impact and management of climate change show that they often lack deep understanding and, in some cases, mirror the types of misconceptions reported in the literature.

Students' understanding and misconceptions

Documented misconceptions in the literature commonly held by students, such as confusing GHE with ozone layer depletion (Francis, Boyes, Qualter, &

Stanisstreet, 1993; Koulaidis & Christidou, 1999; Lee, Lester, Ma, Lambert, & Jean-Baptiste, 2007; Cordero, Todd, & Abellera, 2008; Hansen, 2010), were also observed in the way Singapore students discourse the phenomenon of climate change. Andersson and Wallin's (2000) explication of how students use the ozone model of explaining the greenhouse phenomena is evident throughout this study. Similar to study respondents in other areas (Boyes, Chuckran, & Stanisstreet, 1993; Francis, Boyes, Qualter, & Stanisstreet, 1993; Gowda, Fox, & Magelky, 1997; Papadimitriou, 2004), students in Singapore also tend to lump together all environmental issues to constitute climate change. However, unlike their counterparts in other countries who confuse lead pollution, radioactive contamination and acid precipitation with climate change, Singapore learners tend to be confused and consider tsunamis, earthquakes and general pollution as caused by the phenomenon. It is also believed that climate change causes skin cancer, a pattern also established in Jeffries, Stanisstreet, and Boyes's (2001) research.

Similar to Henry's (2000) respondents, some participants in the study believe that the effects of climate change will be sudden and catastrophic, albeit not discernible at present. In Henry's study, it was found that the respondents were not familiar with anthropogenic contributions to climate change. In contrast, this study shows that Singaporean students are aware of human-induced changes but are generally uninformed about the natural causes of climate change. Moreover, this awareness does not necessarily translate to a clear understanding of its causes, consequences and strategies to manage the phenomenon, a pattern that has been observed by Pruneau et al. (2001).

Sources of misconceptions

It is clear to the greatest number of students (57.7%) why climate change is taught in the classroom. Whether they are exposed to alternative modes of learning, 35.7% answered that they know of such activities while 25.2% stated the opposite. The media again figured as the most pervasive mode of learning about climate change (55.9%) over classroom discussions. Additionally, 57.9% were uncertain whether they find it difficult to discern when climate change is discussed in school. It must be noted that for all of the items stated, the neutral option gathered quite a number of responses indicating that they may not be aware that climate change has been taught through the formal curricular subjects (Table 4.7).

When asked about where they learn most of their climate change information from, the top sources identified were the media (Internet, 75.8%; television, 74.8%; and newspapers, 62.5%), school (74.3%) and popular events such as the Earth Hour (71.5%). Of lesser prominence were government, non-government organisations/community councils and activities/events initiated by the Science Museum, organise events. While the findings show an almost equivalent proportion of students who claim to learn about climate change from school and mass media (about three quarters in each case),

Table 4.7 Singapore students' perception of climate change education in the formal curriculum

	Strongly agree	*Agree*	*Neutral*	*Disagree*	*Strongly disagree*	*No answer*
It is not clear to me why climate change should be taught in schools	11 (2.8%)	31 (7.8%)	113 (28.5%)	172 (43.3%)	57 (14.4%)	13 (3.3%)
We are exposed to alternative modes of learning outside of the classroom (camps, training)	37 (9.3%)	105 (26.4%)	142 (35.8%)	59 (14.9%)	41 (10.3%)	13 (3.3%)
I learn more about climate change through television shows and the Internet than from school discussions	66 (16.6%)	156 (39.3%)	124 (31.2%)	26 (6.5%)	11 (2.8%)	14 (3.5%)
I find it difficult to discern when climate change is discussed in school as it is always integrated with other topics	15 (3.8%)	51 (12.8%)	230 (57.9%)	65 (16.4%)	23 (5.8%)	13 (3.3%)

more than half of the respondents agreed that "I learn more about climate change through television shows and the internet than from school discussions" as shown in the earlier Table 4.7 (Table 4.8).

Attitudes and behaviour

The students reported feeling fortunate that they are not experiencing calamities that beset other countries owing to the belief that Singapore is sheltered from these disasters. Responses such as the following were common:

> We've never experienced droughts and tsunamis or severe storms or typhoons or hurricanes. So, in a sense, we don't feel the effects, so we're very complacent that oh, since it's not here, then we don't need to be very concerned about it.

Indeed the survey determined that almost half of the sample (47.3%) said they enjoyed learning about climate change in school while 41.3% were hesitant to agree or disagree with the statement. Whether they lose interest whenever the

Table 4.8 Sources of information on climate change for Singapore students

Source	*Strongly agree*	*Agree*	*Neutral*	*Disagree*	*Strongly disagree*	*No answer*
School	119 (30.0%)	176 (44.3%)	62 (15.6%)	14 (3.5%)	13 (3.3%)	13 (3.3%)
Television	121 (30.5%)	176 (44.3%)	69 (17.4%)	9 (2.3%)	9 (2.3%)	13 (3.3%)
Internet	129 (32.5%)	172 (43.3%)	58 (14.6%)	13 (3.3%)	12 (3.0%)	13 (3.3%)
Newspapers	90 (22.7%)	158 (39.8%)	94 (23.7%)	29 (7.3%)	11 (2.8%)	16 (3.6%)
Events organised by the government	28 (7.1%)	73 (18.4%)	177 (44.6%)	65 (16.4%)	36 (9.1%)	18 (4.5%)
Events organised by groups, community councils	24 (6.0%)	68 (17.1%)	176 (44.3%)	73 (18.4%)	36 (9.1%)	20 (5.0%)
The Science Museum	41 (10.3%)	79 (19.9%)	154 (38.8%)	68 (17.1%)	36 (9.1%)	19 (4.8%)
Popular global events (e.g., Earth Hour)	149 (37.5%)	135 (34.0%)	73 (18.4%)	11 (2.8%)	13 (3.3%)	16 (4.0%)

Table 4.9 Singapore students' attitude towards climate change education

	Strongly agree	*Agree*	*Neutral*	*Disagree*	*Strongly disagree*	*No answer*
I particularly enjoy learning about climate change in school	47 (11.8%)	141 (35.5%)	164 (41.3%)	14 (3.5%)	16 (4.0%)	14 (3.5%)
When climate change issues are discussed in school, I tend to lose interest	15 (3.8%)	34 (8.6%)	130 (32.7%)	152 (38.3%)	53 (13.4%)	13 (3.3%)

topic comes up, 51.7% expressed disagreement while 32.7% stayed neutral; 12.4% said they do become disinterested in the topic (Table 4.9).

As to whether they are interested in learning more about climate change, the greatest number (48.6%) claimed to be inclined to learn more while 38.8% remained ambivalent about it. This presents an interesting observation that while they claim to enjoy learning about climate change, they are reluctant to learn more about it.

Questions on students' behaviour towards learning about climate change elicited varied responses. On the question of whether students study climate change to gain good marks in the assessment, the most number (45.3%) said they are uncertain while 31.5% admitted to doing so. Some (19.4%)

Table 4.10 Singapore student's behaviour in relation to climate change education

	Strongly agree	*Agree*	*Neutral*	*Disagree*	*Strongly disagree*	*No answer*
I study climate change topics to gain good marks in the assessments	37 (9.3%)	88 (22.2%)	180 (45.3%)	56 (14.1%)	21 (5.3%)	14 (3.5%)
I spend as much time learning about climate change issues compared to other topics	14 (3.5%)	50 (12.6%)	178 (44.8%)	107 (27.0%)	33 (8.3%)	14 (3.5%)
I am a member of an environmental group/interest group whose advocacy is related to climate change	15 (3.8%)	26 (6.5%)	113 (28.5%)	151 (38.0%)	77 (19.4%)	15 (3.8%)

disagreed with the statement. Whether they spend as much time learning about climate change as with other topics, 44.8% expressed ambivalence while 35.3% disagreed with the idea. More than half (57.4%) answered that they are not involved in any advocacy group for climate change (Table 4.10).

Those interviewed expressed their self-doubt about the extent and depth of stock knowledge they hold about climate change. All of the interviewees claim that there is an absence of in-depth understanding. They may have knowledge about the basics of climate change, but it is only on a very shallow level, like one informant called "scratching half of the surface", a trend that can be corroborated by the responses from the survey. Indeed, the majority of the responses show that the students do not have enough knowledge about the topic to have any opinion on the topic (Table 4.11).

Most of the students reported that the effects of climate change are a concern (83.7%). The majority (59.2%) also related that it has personal

Table 4.11 Singapore students' self-perception of knowledge

	Strongly agree	*Agree*	*Neutral*	*Disagree*	*Strongly disagree*	*No answer*
I have learnt a lot about climate change before receiving this survey	29 (7.3%)	107 (27.0%)	179 (45.1%)	58 (14.6%)	22 (5.5%)	2 (0.5%)
I think I have enough information on climate change to have an opinion about it	20 (5.0%)	98 (24.7%)	165 (41.6%)	87 (21.9%)	23 (5.8%)	4 (1.0%)

relevance to them. Some expressed reservation (33.5%) about whether the matter has any bearing on their personal lives, with 6.5% saying it is not a personal concern at all.

Most (68.8%) stated that people could play a part in climate change issues, with 18.4% maintaining a neutral stance and 12.1% stating that people cannot do anything about the phenomenon.

More than majority (55.7%) claimed that turning the air conditioner's temperature a little bit higher will help combat climate change. It is observed that quite a number (28.2%) have reservations about the statement, opting for a neutral position. Moreover, a good 14.6% are not convinced that such an act will help the situation (Table 4.12).

Questions to ascertain reported behaviour of the respondents on climate change have predominantly drawn-out answers that congregate to a Neutral stance. For example, the students are noted to be largely indifferent to encouraging their friends to practice climate change mitigation strategies (44.8%). While 34.0% reported having tried to influence their peers in the past, a good 20.4% admitted to have not been part of any such activity.

They were less indifferent, however, in doing their part to alleviate the problem. A good 72.6% reported being concerned enough to turn off the lights when leaving a room while 14.9% were rather uncertain.

The greatest number of respondents (49.6%) reported being consciously reducing energy usage, while 37.6% were unsure. One in ten (10.6%) showed disagreement, however with the statement.

For the most part, students are inclined to stay Neutral to the statement asking whether they have engaged in activities related to climate change.

Table 4.12 Singapore students' attitude towards climate change

	Strongly agree	*Agree*	*Neutral*	*Disagree*	*Strongly disagree*	*No answer*
I am worried about the effects of climate change	165 (41.6%)	167 (42.1%)	51 (12.8%)	3 (0.8%)	5 (1.3%)	6 (1.5%)
Climate change is a relevant issue to me personally	77 (19.4%)	158 (39.8%)	133 (33.5%)	20 (5.0%)	6 (1.5%)	3 (0.8%)
People could not do anything about climate change because it is a natural occurrence	13 (3.3%)	35 (8.8%)	73 (18.4%)	154 (38.8%)	119 (30.0%)	3 (0.8%)
I do not believe that turning the AC temperature up to 25°C will make any difference in combating climate change	18 (4.5%)	40 (10.1%)	112 (28.2%)	133 (33.5%)	88 (22.2%)	6 (1.5%)

Between the two opposites, those agreeing to the statement (25.4%) slightly outnumbered those who disagree (22.1%). Table 4.13 summarises students' state of understanding both cognitively and affectively about climate change.

Table 4.13 Summary of student readiness

	Causes of climate change	*Impact of climate change*	*Managing climate change*
Knowledge • Recall	30.5% can state what the GHE is	**95.5%** say temperatures will increase	39.3% were able to state one or more mitigation strategies with specific examples given
• Understand	6.3% are able to differentiate between enhanced and natural GHE	**54.9%** were able to discuss the concepts of temperature increase as a result of the enhanced GHE that then leads to melting of the ice caps and then sea level rise	6.5% were able to state and explain mitigation strategies with specific examples given
Skills • Analyse • Evaluate	18.4% were able to draw a relationship between the increase in temperature and the increase in greenhouse gases through analysing the IPCC graphs	5.8% were able to use the information on a hazards map and tried to explain how the impact was caused	1.5% were able to state, describe and explain response strategies with examples and evaluate the usefulness of stated strategies
Values	**52.6%** believe that human activities cause global warming	**92.2%** are convinced that there will not be fewer typhoons	**70.3%** either believed or disbelieved that turning the AC to 25°C will make a difference in combating climate change. The rest had no stands
Action	10.3% are members of an environmental group/interest group whose advocacy is related to climate change		25.4% have done something about climate change

Note: Values greater than 50% have been highlighted in bold.

In terms of student readiness, the study has found that students have relatively strong knowledge about climate change impacts, in no small part, thanks to the ubiquity of information from mass media but are relatively less well informed about the causes and management of climate change. The skills required to discourse climate change are generally lacking although the students have formed their own opinions about the issue. Interestingly, there is a reported concern about taking action for climate change although a small percentage have claimed they have done something about climate change. Perhaps the inaction and the lack of deep understanding are not independent.

The literature on CCE has also found that while there is high general awareness about climate change, there also exists a wide chasm between mere awareness and meaningful understanding of it. This is observed among laypeople (Henry, 2000; McBean & Hengeveld, 2000; Pruneau et al., 2001), teachers (Fortner, 2001; Padpadimitriou, 2004) and students (e.g., Andersson & Wallin, 2000; Lee, Lester, Ma, Lambert, & Jean-Baptiste, 2007) alike. Taking heed from Meinhold and Malkus (2005) on the importance of attitude in gauging environmental behaviour, there is a need to examine the intricacies of young learners' attitudes via the dimensions of their general views towards climate change and climate change learning.

The gap between attitudes and behaviour

Against the backdrop of indifferent attitude to and missed learning opportunities about environmental issues in the case of the Singapore students, a further analysis to determine the correlation, if any, between students' attitude towards climate change learning and their general attitude towards climate change was undertaken.

While youths see climate change as a very serious (Lowe, 2006; Special Eurobarometer, 2009; CEED Consulting, 2011), potentially catastrophic, but distant global problem (Lowe, 2006) that threatens the very future they will be in charge of, around the world, young people profess to be ready and willing to act in mitigating the effects of a changing global climate.

Young people are optimistic that climate change is not a problem without cure. They disagree with the thought that it is an unstoppable process (Special Eurobarometer, 2009) and are particularly confident of their knowledge about the phenomenon (CEED Consulting, 2011). For instance, in a European poll, students along with managers and white-collared workers are among those who claim to be most aware of climate change as an environmental occurrence (Special Eurobarometer, 2009).

Certainly, for a meaningful comprehension of how climate change attitudes are formed, one must consider the context. While climate change is truly a global phenomenon, it is perceived, interpreted and understood differently in diverse settings as the discourse gets indigenised. Thus, in examining how Singaporean youths approach the issue, it is important to look at the matter from multiple perspectives.

To this end, factor analysis was performed on the 16 statements that were retained from the original 24 statements of the same study cited above. For items phrased in the negative, computations were employed both for actual scores and reversed scores, after which the results of the factor analyses were compared. The items were noted to be consistent in their loading for both instances. The reported factors in this study used actual scores. Cleaning the data involved eliminating entries with no answer, thus decreasing the original number of respondents to 366, a 92.19% of the original count. Therefore, for the rest of the discussion, the N = 366 is used.

The data derived were subjected to Principal Components Analysis with Varimax rotation. A two-factor solution, which explains 55.109% of the total variance, was regarded as the most parsimonious grouping in that factors generated were composed of items that hang together numerically and logically. Items with factor loading ≥ 0.40 per factor were the ones considered for analysis.

The different constructs derived through factor analysis are so labelled and defined according to the items that constitute the resulting subscales. Hence, *engagement* in learning about climate change refers to a student's motivation to learn about the subject. On the other hand, *disengagement* indicates a learner's confusion and lack of interest in the topic. The construct *knowledge perception* refers to a person's confidence in the amount and extent of information he/she possesses to make an informed opinion about climate change. *Involvement* pertains to the sense of ownership one has on the matter, whether it is of relevance to him/her and whether he/she sees himself/herself as an active element of the solution. *Non-involvement* points in the opposite direction, mainly alluding to an individual's propensity to be dismissive and unaffected about the issue.

The items that comprise each factor and the associated Cronbach's alpha values are as follows (Table 4.14).

In addition, a three-factor solution for general climate change attitude, which explains 61.083% of the total variance, was derived. Similarly, a factor loading ≥ 0.40 was considered for each of the items. The resultant factors are as given in Table 4.15).

The resultant factors are observed to be pointing to polarised attitudes towards climate change learning and also towards general disposition towards the phenomenon. The three broad dimensions uncovered from the factor analysis are engagement with the issue, self-knowledge and involvement with the issue.

Engagement in learning about climate change

On the question of whether students study climate change to gain good marks in the tests or examinations, 47.0% said they are uncertain while 33.1% admitted to doing so. Some (20.0%) disagreed with the statement. This indicates that while extrinsic motivation is the key factor in student engagement

Table 4.14 Two-factor solution for principal component analysis of students' attitudes

Factors	*Cronbach's alpha*
Engagement with climate change learning	**0.666**
1. I particularly enjoy learning about climate change in school	
2. I study climate change topics to gain good marks in the assessments	
3. I want to gain a deeper and more meaningful understanding of the mechanics and concepts of climate change	
4. I learn more about climate change through television shows and the Internet than from school discussions	
Disengagement with climate change learning	**0.726**
1. It is not clear to me why climate change should be taught in schools	
2. When climate change issues are discussed in school, I tend to lose interest	
3. I find it difficult to discern when climate change is discussed in school as it is always integrated with other topics	

in Geography learning in Singapore, this is not the case for CCE. Indeed, the disengagement reported in some items needs to be explained by factors other than extrinsic motivation.

Further, almost half of the sample (49.2%) say they enjoy learning about climate change in school while 43.2% are rather hesitant to agree or disagree with the statement. As to whether they are interested in learning more about climate change, the greatest number (50.2%) claim to be inclined to learn more while 40.4% remained unsure about it. The media turn out to be the more pervasive mode of (57.4%) over classroom discussions.

The subscale mean score (m = 2.541) shows that the students are predominantly motivated, but not decidedly so, to learn about climate change. This is not dissimilar to youth around the world where young people claim that topics covering global warming, its causes and effects, interest them (CEED Consulting, 2011). Maintaining cognitive engagement, however, appears trickier. Mower (2012) illustrated in his research that despite young people's significantly high level of awareness about the discourse, they continue to be cognitively disengaged and have a consistently low feeling of responsibility on the issue.

Disengagement towards climate change learning

It is clear to the greatest number of students (59.9%) why climate change is taught in the classroom. Whether they lose interest whenever the topic

Table 4.15 Three-factor solution for principal component analysis of students' attitudes

Factors	*Cronbach's alpha*
Perception of one's own knowledge about climate change issues	**0.738**
1. I have learnt a lot about climate change before receiving this survey	
2. I think I have enough information on climate change to have an opinion about it	
Involvement and ownership in mitigating climate change	**0.675**
1. Climate change is a relevant issue to me personally	
2. I have encouraged my friends to practice climate change mitigation strategies	
3. I am consciously reducing my energy usage in order to lessen my contribution to hastening climate change	
4. I take part in activities that rally for greater awareness about climate change	
Non-involvement in mitigating climate change	**0.675**
1. I go on with my life as usual. I cannot be bothered turning off the lights when I leave a room	
2. People could not do anything about climate change because it is a natural occurrence	
3. I do not believe that turning the AC temperature up to 25°C will make any difference in combating climate change	

comes up, 52.5% expressed disagreement while some (34.7%) stayed neutral; 12.8% said they do become uninterested in the topic. Additionally, 60.1% are uncertain whether they find it difficult to discern when climate change is discussed in school.

The subscale mean score shows that students' responses cluster just above the Likert midpoint (m = 3.396), suggesting that while students tend to disagree with the negative statements, they are also drawn to adopting a neutral stance on whether climate change as a topic for learning engages them effectively.

Self-assessment of own knowledge

The first item asks for the student's perception about the amount of learning he/she has on climate change. When computing for agreement, 34.7% appear to concur with the statement while those who think otherwise are at 19.7%.

The second question probes further by asking the students to evaluate whether the knowledge they hold qualifies them to make a strong point on the matter. Agreement is pegged at 30.0% while non-agreement is at 27.6%.

The students' scores congregate towards the neutral position for both items. The questions were asked, to elicit answers pertaining to the students' confidence in their knowledge stock about climate change and whether they perceive to have enough or lack it.

While the percentage for agreement is higher than non-agreement, the latter should not be ignored indiscriminately but must be analysed vis-à-vis the more prevalent "Neutral" position (45.6% and 42.3%, respectively). Further, the subscale score of m = 2.9098 confirms that students do not have much confidence that what they know about climate change – whether it be learnt in school or elsewhere – would collectively make a good enough knowledge stock to enable them to have an informed opinion about the matter. The neutral option is an attractive alternative to taking a definite stand on the issue.

Involvement and ownership in climate change mitigation

The majority (59.5%) of the respondents agreed that climate change has personal relevance to them. Some expressed reservation (34.4%) about whether the matter has any bearing on their personal lives, with 6.0% saying it is not a personal concern at all.

The students are largely indifferent to encouraging their friends to practice climate change mitigation strategies (45.9%). While 33.9% reported having done so, a good 20.2% also expressed otherwise. This is in contrast to other studies around the world where youth are quick to express their readiness to undertake activities that will contribute to combating its exacerbation. In some studies around the world, the youth are willing to encourage their inner circle, family and friends, to act responsibly for the environment (Kuo, 2010; CEED Consulting, 2011) as well as support government policies aimed at curbing greenhouse gas emissions (Kuo, 2010).

The greatest number of respondents (49.1%) reported being consciously reducing energy usage, while 39.9% are rather unsure. For the most part, students are inclined to stay neutral to the statement asking whether they have engaged in activities related to climate change. Between the two opposites, those agreeing with the statement (26.0%) slightly outnumbered those who disagreed (21.3%). The subscale mean scores for the items indicate that students' responses clustered between 2.2869 and 2.9317 (m = 2.64755), indicating that the respondents tend to agree with the items but are also bent towards adopting a neutral stance.

Non-involvement in mitigating climate change

The students professed to be not indifferent in doing their part to alleviate the problem. Close to three in four (73.5%) reported being concerned enough to turn off the lights when leaving a room while 14.8% were unsure about it. In addition, most (69.7%) believe that others could play a part in climate change issues, with 18.6% maintaining a neutral stance and 11.7% stating that people cannot do anything about the phenomenon.

More than majority (56.3%) believe that turning the air conditioner's thermostat a bit higher will help combat climate change. It is observed that quite a number (29.2%) have reservations about the statement, opting for a neutral position. Moreover, a good 14.5% do not believe that such an act will help the situation. The subscale mean score illustrates that students' responses cluster above the midpoint of the 5-point Likert Scale (m = 3.806). This pattern indicates that the respondents were slightly more inclined towards having a neutral attitude on the issue, while predisposed to disagreeing with the negative statements.

In all, while students in Singapore reported that they have done something about climate change, there is still a substantial group of them that do not believe that anything can be done about climate change. In studies around the world, youths do not possess a strong belief that their efforts can positively influence the big picture of environmental stewardship despite their confidence, optimism and concern (Schreiner, Henriksen, & Hansen, 2005; Ahmad, Rahim, Pawanteh, & Ahmad, 2012). They retreat to an attitude of non-involvement once the discourse shifts towards personal responsibility over climate change management. They then become predisposed to claim that other parties are to be blamed for the problem (Kuo, 2010; CEED Consulting, 2011).

On the other hand, they may acknowledge their part in the creation of the problem (Mower, 2012) and even recognise the significance of coming up with a solution (Kuo, 2010; Ahmad, Rahim, Pawanteh, & Ahmad, 2012). In spite of this, they continue to have low concern for the consequences of changes in the climate (Mower, 2012), refusing to take personal action (Special Eurobarometer, 2009) and dismissing the phenomenon as a difficult and overwhelming matter that is beyond the control of a young person to manage (Kuo, 2010). Finally, Lowe (2006) posited that the way climate change is framed – as a distant, perilous and intangible event – is becoming an annoying and unwanted subject for young people.

Correlations between factors

How the factors relate to one another, was then determined statistically by using the subscale mean scores from the survey. Table 4.16 shows the Pearson correlation coefficients.

Engagement with learning and involvement with mitigating climate change are moderately and positively related ($r = 0.512, p < 0.001$). The data shows that those who possess a positive attitude towards climate change are also more interested to learn more about it and *vice versa*. Also, engagement towards learning is negatively and moderately correlated with a student's own knowledge perception ($r = -0.379, p < 0.001$).

This result shows that students who claim to know a lot about climate change are less motivated to keep learning more about it. Perhaps this is an indication of confidence in one's knowledge about climate change that led to students' general belief that they need not learn more. Unsurprisingly,

Table 4.16 Correlation coefficients between factors

	Disengagement (Learning)	*Involvement (Issue)*	*Non-involvement (Issue)*	*Knowledge perception*
Engagement (Learning)	0.054	0.512***	−0.030	−0.379***
Disengagement (Learning)		−0.145**	0.416***	0.079
Involvement (Issue)			−0.169***	−0.316***
Non-involvement (Issue)				0.039

** Correlation is significant at 0.01 level (2-tailed).
*** Correlation is significant at 0.001 level (2-tailed).

there is no observed correlation between the bipolar scales of engagement and disengagement and between engagement and non-involvement in mitigating climate change.

On the other hand, the attitude of disengagement towards learning and one's propensity to be involved in mitigating climate change are weakly and negatively related ($r = -0.145$, $p = 0.006$). Those who do not see climate change as a topic that interests them may also be likely to be uninvolved in mitigation advocacies. Most logically, non-involvement and disengagement to learn about climate change are moderately, positively related ($r = 0.416$, $p < 0.001$). Students who do not see the sense in learning about climate change are more likely to be unaffected and dismissive of efforts to mitigate its effects.

There is no significant relationship observed between disengagement in learning and knowledge perception.

A negative, albeit weak, relationship is observed between attitudes of involvement and non-involvement in mitigating climate change ($r = -0.169$, $p = 0.001$). Some students who view climate change to be a relevant issue in their lives are less likely to be indifferent about it and *vice versa*. In addition, an inverse relationship is observed between students' perception of the extent of their own knowledge about climate change *vis-à-vis* attitude of involvement ($r = -316$, $p < 0.001$). This means that there are fewer of those who are confident with their perception of climate change knowledge, who actually claim to be actively involved in the climate change issue.

On the other hand, there is no observed correlation between students' perception of their own knowledge and their tendency to be uninvolved with the issue of climate change.

By analysing the relationship between the factors, it becomes clear that students who are more motivated to learn about climate change are those that feel more involvement and ownership in mitigating climate change and those who feel that they do not have enough knowledge about the issue. Indeed, the correlation between involvement with mitigating climate change and motivation to learn as well as between non-involvement and disengagement to learn about climate change are both positively related with considerable statistical significance. In other words, both the involvement–engagement relationship and the corollary of the non-involvement–disengagement relationship are true. While this finding reaffirms the role of CCE as providing the necessary knowledge to close the gap between attitude and action, or engagement and involvement in this case, it is perhaps interesting to examine youths' propensity for engagement and involvement within the larger social context.

Trumper (2010) argued that the context upon which attitudes are formed is important in truly understanding the intricacies of how young people approach climate change. The author illustrated in his research how a country's level of development has a strong relationship with students' interest in learning about environmental issues. According to Trumper (2010), students from highly developed economies tend to dismiss environmental problems as overstated. They think it is the responsibility of other parties to come up with solutions, are hardly involved and are less willing to compromise personal comforts. On the other hand, young people from countries of medium and low development appear to be more concerned, optimistic and involved in environmental issues at a personal and collective level. They are also observed to have more interest to learn about environmental topics than do their counterparts in the First World, even to the point of considering a career in environment protection.

The context of environmentalism in Singapore

Environmentalism in the city-state was initially in the early years of independence, as a comprehensive scheme by the government to create a garden city (Kong et al., 2000). This was to remedy the depletion of its natural landscape that has been replaced by rapid urbanisation for industrialisation and economic development. In its place developed a different kind of nature, called "managed messicol vegetation", with which most of the present-day Singaporeans have become familiar with (Kong, Yuen, Sodhi, & Briffett, 1999, p. 323).

Much like many policies implemented in Singapore, environmentalism is pursued with the rationality of developmental gain sans ethical impetus (Kong et al., 2000; Neo, 2007). Teo (2004) highlighted the political dimension to the discourse in that environmentalism evolved into a tool for social engineering, for Singapore citizens:

> to feel a sense of pride and belonging to their country because it has a clean, safe and pleasant environment, a pride which would help cohere Singapore's pluralistic populace to achieve a culture of solidarity which will, in turn, boost investor confidence in the economy; at the same time, the government wants Singaporeans to care for Singapore's physical environment by making them believe that Singapore belongs to them.
>
> (Teo, 2004, p. 16)

Affluence and modernisation also have influence over the values and attitudes of the population towards nature and the environment. Living in a regional hub of commerce, trade and culture, Singaporeans are epitomised to be the materialistic and pragmatic urbanites (Jiuan, Wirtz, Junk, & Keng, 2001), with hardly any experience in their own country with nature in its unmanaged, undisturbed state (Kong, Yuen, Sodhi, & Briffett, 1999) and driven to acquire wealth in order to attain success (Jiuan, Wirtz, Junk, & Keng, 2001). While they are aware of environmental problems and are keen to help in preserving nature, this comes with the condition that they do not have to sacrifice personal comfort and benefits in the process (Figueres, 2012). This unidirectional transference of responsibility is based on causality rather than morality. They generally believe that large corporations are responsible for polluting the environment and must therefore be solely accountable to clean it up (Jiuan, Wirtz, Junk, & Keng, 2001).

Pragmatism still figures as the key determinant for environmental behaviour in Singapore such that it is pursued only if gains are apparent. For example, recycling is practised if it proves to be cost-effective to do so. Thus, it is unlikely that any project on environmental management be effectively anchored on citizens' individual conscience but may need the strong hands of the state for it to be sustained (Kong et al., 2000).

Growing up immersed in this scene of affluence, statecraft and urbanity, the youths in Singapore have unsurprisingly little interest and affinity with nature demonstrating an attitude that is a result of their limited contact with nature, of overprotective parents as well as of exposure to a plethora of recreational and entertainment choices (Kong, Yuen, Sodhi, & Briffett, 1999). It is thus a paradox that environmental awareness is high in Singapore students' mindsets according to the National Environment Agency (National Environment Agency, 2006).

In fact, they view ozone depletion and global warming as environmental problems that they are most concerned about (Tan, Lee, & Goh, 1998). Despite their cognisance of these issues, they are unwilling participants to come up with solutions. They are neither keen to volunteer their free time for environmental advocacies nor are they agreeable to buying recycled products at a higher cost. Just like their adult kinfolks, many young people believe that the government should be responsible to keep the environment clean and green and that they have more important things to attend to than environmental issues (National Environment Agency, 2006). It is perhaps this socially and culturally embedded indifferent stance on environmental concerns that has shaped the way youths view climate change in Singapore. If the social climate has not provided a nurturing condition for youths to develop an affinity for nature, then environmental knowledge can be a viable alternative to providing a rationale for action, at least drawing on the concept of causality – personal responsibility in contributing to the climate change issue.

This is affirmed by the results of the factor analysis in this study where which reaffirms the role of CCE as providing the necessary knowledge to close the gap between engagement and involvement. Perhaps more fundamental was the disjoint between what they know and what they can do. An argument can be made that if students have accurate knowledge about the issue there is empowerment and hence more ownership of the behavioural change that they can take up.

Correcting student's misconceptions about climate change

'The Hole in the Sky Causes Global Warming,' a 2016 study, attempts to determine whether classroom instruction can successfully correct learners' misconceptions about climate change, thereby improving the knowledge base from which they can make decisions about behavioural change. At the same time, the study intends to highlight the resilience of misconceptions. Classroom instruction was delivered through non-direct refutation and took the form of a teacher-led pedagogy without researcher influence. Many studies (Table 4.17) have established the existence of misconceptions in CCE research, but there has been less attention paid to questions about how resilient misconceptions are to classroom instruction.

Chang and Pascua (2016) used a case study technique to diagnose misunderstandings. While other methods for determining misunderstandings have been used (Boon, 2010; Punter, Ochando-Pardo, & Gracia, 2011; McNeill & Vaughn, 2012), this study used a diagnostic exam modelled after Treagust's (1988) two-tier methodology. The tiered test is divided into two sections: the Content Tier, which assesses subject matter knowledge, and the Reason Tier, which examines the supporting concepts that underpin content knowledge. The items are specially designed to detect misconceptions in climate change learning using the multiple-choice question (MCQ) format. This two-tiered test provides insight into students' reasoning processes that could not This two-tiered test gives information on students' thinking processes that could not be obtained through other means. For example, it enables researchers to examine how students reason using both domain-specific and domain-general information (Tsui & Treagust, 2010).

In the first session, the instructor began the intervention program by introducing the issue of climate change by presenting the film *The Day after Tomorrow*. Students were asked to summarise the plot of the movie, state possible causes of climate change as depicted in the video, argue whether the scenario depicted in the video is likely to happen in the future and state their ideas about the consequences of climate change in a worksheet distributed the following day. The students were given ten minutes to answer the questions before some students were called out to present their responses for points. Throughout the following sessions, this teaching technique was employed consistently. On the last day, the students were divided into groups of two or three to report on the causes, impacts and management of climate change.

Misconceptions can arise from the incorrectness, incoherence and incompleteness of students' mental models concerning climate change. The

Table 4.17 Misconceptions identified in the literature

Misconceptions	*Literature cited*
Unable to identify most of the greenhouse gases; believes CFCs are the key greenhouse gases responsible for anthropogenic climate change	(Boon, 2010; Hansen, 2010; Bozdogan, 2011; Liarakou, Athanasiadis, & Gavrilakis, 2011; Punter, Ochando-Pardo, & Gracia, 2011; McCuin, Hayhoe, & Hayhoe, 2014; Dawson & Carson, 2013; Bodzin et al., 2014; Chang & Pascua, 2016)
Erroneous conceptions of the causes of global warming (e.g., nuclear power, solar energy, natural gas not a fossil fuel)	(Boylan, 2008; Coşkun & Aydın, 2011; Liarakou, Athanasiadis, & Gavrilakis, 2011; Ocal, Kisoglu, Alas, & Gurbuz, 2011; Yazdanparast et al., 2013)
"The hole in the sky is causing global warming" Conflation of ideas about the ozone layer and the GHE	(Lee, Lester, Ma, Lambert, & Jean-Baptiste, 2007; Boon, 2010; Hansen, 2010; Çimer, Çimer, & Ursava, 2011; Liarakou, Athanasiadis, & Gavrilakis, 2011; Ocal, Kisoglu, Alas, & Gurbuz, 2011; Punter, Ochando-Pardo, & Gracia, 2011; Arslan, Cigdemoglu, & Moseley, 2012; Ikonomidis, Papanastasiou, Melas, & Avgoloupis, 2012; Lambert, Lindgren, & Bleicher, 2012; Mower, 2012; Dawson & Carson, 2013; Chang & Pascua, 2016)
Unable to differentiate the natural and enhanced GHE	(Lee, Lester, Ma, Lambert, & Jean-Baptiste, 2007; Boon, 2010; Liarakou, Athanasiadis, & Gavrilakis, 2011; Ocal, Kisoglu, Alas, & Gurbuz, 2011; Arslan, Cigdemoglu, & Moseley, 2012; Dawson & Carson, 2013; Chang & Pascua, 2016)
Climate change is believed to be solely a human-induced phenomenon; no understanding of the natural causes of climate change	(Çimer, Çimer, & Ursava, 2011; Bodzin et al., 2014; Chang & Pascua, 2016)
Confusion on the effects of climate change (e.g., the sea level will stay the same, the effects are uniform, hotter climate every single year, skin cancer, earthquakes, tsunamis, acid rain)	(Lee, Lester, Ma, Lambert, & Jean-Baptiste, 2007; Shepardson, Niyogi, Choi, & Charusombat, 2009; Boon, 2010; Coşkun & Aydın, 2011; Liarakou, Athanasiadis, & Gavrilakis, 2011; Ocal, Kisoglu, Alas, & Gurbuz, 2011; Arslan, Cigdemoglu, & Moseley, 2012; Mower, 2012; Yazdanparast et al., 2013; Chang & Pascua, 2016)
Unfamiliarity with mitigation and adaptation strategies (e.g., the Kyoto Protocol is confused with the Montreal Protocol, "good things" like clean beaches can help ameliorate the enhanced GHE, the use of unleaded gasoline would help reduce global warming)	(Boyes, Stanisstreet, & Yongling, 2008; Liarakou, Athanasiadis, & Gavrilakis, 2011; Arslan, Cigdemoglu, & Moseley, 2012; Ikonomidis, Papanastasiou, Melas, & Avgoloupis, 2012; Yazdanparast et al., 2013)

(*Continued*)

Table 4.17 (Continued)

Misconceptions	*Literature cited*
Global climate change means pollution in general	(Boyes et al., 2008; Shepardson, Niyogi, Choi, & Charusombat, 2009; Punter, Ochando-Pardo, & Gracia, 2011)
Ultraviolet radiation is the culprit in global warming	(Çimer, Çimer, & Ursava, 2011; Dawson & Carson, 2013)

pre-test, for example, showed that students had many misconceptions, with many only able to properly answer the Content Tier, resulting in many instances of mixing of incorrect and accurate notions. Consistent in the literature, the misconceptions revolved around a lack of understanding of climate change as both a natural and a man-made phenomenon, the natural GHE and its properties, the enhancement of the GHE, the elements involved in heat-trapping and their characteristics Furthermore, there were significant misconceptions in the students' understanding of the effects of climate change, particularly how it relates to other non-atmospheric events such as tsunamis, earthquakes, acid rain and skin cancer. There is also confusion between the Montreal and Kyoto Protocols as the primary treaty aimed at curbing greenhouse gas emissions. The post-test results showed that, while there was substantial progress in knowledge as evidenced by the rejection of some misconceptions, the majority of them were preserved, albeit in smaller numbers

This study backs up Von Aufschnaiter and Rogge's (2010) hypothesis that poor understanding is caused mostly by missing concepts and a lack of explanatory conceptual understanding available to learners. At the grain size of a single idea, inconsistencies in the form of incomplete conceptualisation may be recognised. Under this condition of prior knowledge, gaps in an imperfect idea can be filled (Chi, 2008). New notions, on the other hand, were hard to incorporate since they were so strongly at odds with existing knowledge. One example is the confusion between infrared, ultraviolet and visible light as the type of radiation that is absorbed and re-emitted. Water vapour, on the other hand, was successfully recognised as a greenhouse gas since it did not have to supplant another notion.

Mental models are produced as internal representations of an external structure when concepts are grouped to express linkages and interconnectedness. A defective mental model employs a coherent but inaccurate mental model that makes various predictions and explanations and may incorporate components from the right model (Chi, 2008). Furthermore, Chi (2013) contends that persistent misconceptions arise from the mis-categorisation of processes as sequential while, in reality, they are emergent. Although the identified misconceptions were shown to be resistant to rebuttal, Chang and Pascua (2016) argue that the models, both the misconceptions and the correct ones, were at the same categorical level (as a sequential process), therefore a category shift was not required. They also claimed that the inability to dispel misconceptions was due to a lack of direct and iterative rebuttal addressing the key assumptions of the accurate climate change model in comparison to the misunderstood models. In some ways, Chang and Pascua's (2016) findings confirm the necessity for correspondingly effective pedagogical practices that could immediately rectify these misconceptions. Indeed, a purposeful shift toward more explicit recognition, assimilation and direct rebuttal of erroneous knowledge systems will be beneficial.

Summary of what students know, believe and do about climate change

In summary, while Singapore students commonly associate events such as droughts, storms and hurricanes with climate change, their conceptualisation is not deep enough to allow them to connect the causes of the phenomenon to its impacts. They are able to identify one or two strategies but are not able to explain or evaluate the usefulness of these strategies. Thus, they are mostly ambivalent towards learning about it in-depth, merely seeing the matter as part of a larger environmental topic on which they will have a test on.

While the students profess to have received a great deal of information on climate change, they do not have the confidence to claim that they hold sufficient knowledge stock to have a strong opinion about it. In testing for the presence of misconceptions reported in the literature, it was found that students do hold erroneous information about climate change. The most prominent of which is the belief that the depletion of the ozone layer contributes to global warming. Other misconceptions include the confusion between greenhouse gases and chlorofluorocarbons, misunderstanding of the UV rays against infrared rays and the belief that all natural disasters such as tsunamis and earthquakes are closely linked to climate change.

While students generally profess a positive attitude towards climate change, they are more hesitant to respond to the call for behaviour change. They are largely indifferent to encouraging their friends, for example, to practice the 3Rs. They themselves are not keen to participate in activities and advocacies that are related to climate change. Most are saying that they are consciously reducing energy usage like turning off the lights when leaving a room, but the interviews reveal that such behaviour could best be attributed to habit and/or parental control.

Among the students' sources of information on climate change, activities comprising the informal curriculum (events organised by the government, non-government organisations/community councils and activities/events initiated by the Science Museum) are least prominent except for the Earth Hour. The media through the Internet, television and newspapers figure as the top source of information, followed by classroom learning.

In a gist, this study highlights the extent to which youths of Singapore engage with the discourse of climate change in both their academic and personal lives. The findings show that teenagers in Singapore, at least those at the

Secondary 2 level, view climate change from a neutral, distanced perspective – a discovery that is not dissimilar to the findings of Lowe (2006). The constructs derived determine that the youth have formed an attitude towards climate change that is distributed along the continuum of involvement and non-involvement. Similarly, a bipolar distribution is seen in how they view learning about climate change, that of engagement and disengagement.

Young people claim to be rather motivated to learn about climate change but are largely undecided on whether the topic is an engaging subject or not. They believe that climate change is relevant to their lives and are not apathetic about it, yet they do not position themselves on either of the extremes (activism-apathy) but are bent towards having a neutral, ambivalent stand. Perception of knowledge is related negatively to both the attitude of involvement and engagement in learning. Indeed, teenagers who believe that they have a part to play in solving the crisis of climate change will be keener to learn about the issue. More importantly, young people who feel they already know enough about the topic will tend to shun away from learning more about it.

Those who reported to be actively involved in the issue of climate change, most logically, are neither dismissive about it nor are disengaged in learning about it. They also tend to be motivated to learn more about the phenomenon. Quite the reverse, students who are indifferent to the issue are also disengaged with the topic in class. Students who do not see the sense in learning about climate change are more likely to nonchalantly breeze through life believing that their actions will not have any effect on the planet. The study provides cues for educators to revisit pedagogies in teaching climate change, focusing on relating the reality and relevance of the issue. The two key implications in motivating students to learn more about the topic are in ascertaining how much they really know about climate change and in bringing inadequacies in knowledge to the foreground, as well as to raise awareness that the issues related to climate change are real and relevant. In addition, there is a need to help youths determine what disinterests them and what are the elements in the discourse that leads them to be ambivalent.

The finding from this mirrors Trumper's (2010) observation that students in developed countries tend to lack concern about environmental issues such as climate change. As is the case in Singapore, the test of education program effectiveness is on how to engage students meaningfully and on how to spur concern strong enough to lead to active involvement. The challenge to engage Singapore's youth cognitively and socially with the discourse remains, considering the socioeconomic and political milieu that learners are exposed to and immersed in every day. What comes to mind was what Trumper (2010) stressed, that educational programs must follow a place-based pedagogy, allowing local and regional narratives to permeate the discourse with due consideration of indigenous culture, education system and economic setting.

In other words, there is an even greater need for systematic education through formal curricula to help learners bridge the gap between knowledge, understanding, beliefs and action. To do so would imply the need for a more contiguous and robust approach to teaching and learning about climate change. To this end, CCE must *help learners develop knowledge, skills, values and action to engage and learn about the causes, impact and management of climate change.* While this chapter has explored the readiness of students to learn about climate change, the same question of teachers' readiness has to be addressed for an effective curriculum to be implemented. Indeed, a curriculum is by itself not just a syllabus or standards document waiting to be enacted, but rather an active interpretation of its content, translated into artful teaching that would lead to effective learning.

5 Pedagogies for successful implementation of CCE

The question of whether the potential of climate change education (CCE) is realisable as a mitigation strategy depends on how well the CCE curricula are carried out in schools. At the classroom level, teachers will have to find the best ways to help students attain the curricular intent – through effective pedagogies. Indeed, schools need to make sure that they have the tools, resources and skills to help teachers educate their students on how climate change will impact them. If they do not do this, then there is no way that CCE can be a viable strategy to help mitigate climate change. There is a need to first find out what sort of pedagogies for CCE is being reported in the literature. While there are different pedagogies one can adopt for the range of learning outcomes on the causes, impact and management of climate change, this chapter will report on empirical findings of successful pedagogies that have been reported in the literature.

Pedagogies for CCE

A basic secondary data analysis approach was undertaken to examine the literature on effective CCE pedagogies. This was achieved by going through a number of scientific journals that best pertain to environmental education in the academic field. They include *Environmental Education Research* (*EER*), *International Research in Geographical and Environmental Education* (*IRGEE*), *Journal of Environmental Education* (*JoEE*) and *Journal of Geography* (*JoG*) for the period 2014–2021.

The articles sourced from the scientific journals mentioned above were analysed at the abstract and article content levels to identify the current state of CCE and to see if there were any changes in trajectory. In order to select the appropriate articles, a filtered search using the words "climate change education" was undertaken. The articles were further filtered and selected based on title relevance. Articles selected hence made up the bulk of secondary data used for the research. Table 5.1 shows how the articles were filtered based on each journal which shows the number of selections for each article.

Articles were specifically selected from these journals based on their SCOPUS rating. In terms of environmental education, the top three

DOI: 10.4324/9781003093800-5

Table 5.1 Articles reviewed in journals of environmental education and geographical education

	Title of journals			
	Environmental Education Research (EER)	*International Research in Geographical and Environmental Education (IRGEE)*	*Journal of Environmental Education (JoEE)*	*Journal of Geography (JoG)*
Number of articles selected	647	289	1790	1680
CiteScore	5.2	3.1	4.2	3.3
Percentile (%) rank in (domain)	94 (Education)	81 (Education)	90 (Education)	78 (Geography and Planning)
Refined to articles based on search word "Climate Change Education"	112	54	89	59

journals are *EER*, *IRGEE* and *JoEE*, with their percentiles in the first tier, that is, in the top 35%. They also have a high cite score, generally higher than the score of 1 within the social science field, representing and giving credibility to the works in environmental education in the academic community. Although *JoG* is not in the top list for SCOPUS under environmental education, the journal was accounted for to further obtain a good sense of possible topics of environmental education in the public domain.

An approach similar to a meta-analysis was conducted to create a form of evaluative and objective standing towards the great diversity of content found within articles. This was achieved with a qualitative analysis using the NVivo software where open coding was executed to form labels and themes of pedagogies. These were performed at the level of not just the title but also on keywords as well as the main body of texts. In order for the coding to work, word clouds were used to provide a visual pattern of what the main clusters of codes were. These were then considered as a master codebook. After all these patterns were examined, a series of themes were derived. These themes can be seen in the schematic in Table 5.2.

Next, a thematic analysis was conducted by coding the secondary data again, where new linkages and relationships were found with a focus on pedagogy. They were then categorised into different levels and broken into different approaches. Table 5.3 illustrates the thematic issues identified for pedagogy.

Table 5.2 General themes identified in the literature review

Who – Teacher, Students or Public/Community
Outcomes or Experiences
Actions and Behaviours
Identity
Perspective and Attitudes
What – Curriculum, Knowledge/Subject Matter or Policies
Curriculum
Sociological Outlook
Citizenship Education
Holistic Development
Knowledge/Subject Matter
Cognitive versus Affective
Policies
Governance
Neoliberalism
Culture
How – Pedagogy or Assessment
Formative and Developmental
Cognitive versus Affective
Linguistics, Language
Authenticity
On site or field-based

Table 5.3 Thematic analysis for issues in pedagogy

Issues on pedagogy	*N (number)*
Hands-On Approaches	10
Conceptual Approaches	17
Project-Based & Place-Based	15
Issues-Based	6
Online & Technology	15
Global & Future Citizenship	10
Teacher Discourse	8
Storytelling	6
Critical-Thinking Approach	11
Disciplinary Approach	27

Lastly, they were reorganised to be reframed around the knowledge, skills, attributes and behaviour outcomes that are expounded below. Further coding was done for easier identification of the themes and how each pedagogy was linked to the knowledge, skills, attitudinal and behavioural learning outcomes for CCE. To a certain extent, all the pedagogies described would target knowledge, skills and attitudinal and behavioural learning outcomes for CCE. Pedagogies that fall under knowledge learning outcomes are approaches that are largely based on understanding and developing cognitive processes and deepening learning. These include pedagogies such as conceptual and disciplinary approaches.

Conceptual approaches range from establishing principles and maxims to help students anchor their cognitive processes – from using notions in climate science like adaption and resilience to designing or arranging instruction, learning progressions and transformative concepts. Transformative concepts, such as using transformative learning and resilient practices, are used as concepts to gain better knowledgeable outcomes. For example, the usage of Participatory GIS projects has transformative impacts on students to think about societal problems and their role as individuals and professionals when addressing these problems (Sinha et al., 2017). Resilient practices taught are used as a concept to encourage action for reflection, where commonalities within "adaptive co-management for resilience contexts include meaningful participation, reflexivity, second-order learning, problem-solving and developing competency" (Krasny, Lundholm, & Plummer, 2010, p. 668). Other approaches such as learning progressions and maxims are utilised to break down complex concepts. Such examples include the use of maxims, where principles serve as a guide for students to gain a more in-depth exploration of the topic where they moved on quickly from simply understanding maxims to appreciating the number of ways they can be applied (Birdsall, 2004, Korson & Kusek, 2016, Block, Goeminne, & Van Poeck, 2018). For learning progressions, one article states how it is "helpful for tackling complex and nebulous concepts and disciplinary domains" (Mohan, 2018, p. 125).

It is important to note that while using conceptual approaches, students and teachers alike have the potential of mis-conceptualising knowledge due to mis-categorisation of concepts. This is often mentioned in misconceptions literature which can prove to be resilient. However, the same conceptual methods are also used to often correct such mis-conceptualisations through the process of reassigning or shifting miscategorised concepts from one ontological category to another ontological category.

On the other hand, disciplinary approaches utilise disciplines that are emphasised in sole disciplines or holistic approaches such as multi-, inter- or cross-disciplinary approaches to instil improved understanding and cognitive development. Articles that focus on sole disciplines mainly advocate for CCE to be taught within that discipline. One article illustrates how the "geography classroom is the best place to teach about climate change as it affords both space and time for students to interact with the topic within disciplinary domain of geography subject" (Chang, 2015, p. 182). There are also those that promote an integrated disciplinary approach, highlighting holistic learning through integrating multiple subject areas. This is witnessed in the "call for integration of environmental and citizenship education and focus more specifically on clarifying how climate change scientific knowledge is framed and positioned within school curricula in relation to broader civic and citizenship concepts and ideas" (Ho & Seow, 2017, p. 241). Through these approaches, students are presented opportunities from different subject areas and knowledge to gain foundations and improvements in understanding.

There are also pedagogies that encourage the development of skills. This includes online and technological approaches which target technological applications and the incorporation of digitalisation into student learning. This group of approaches can be further split into three categories. The first category is when technology is used as a support for learning. Digital tools such as digital mapping and geospatial technology are introduced and posited to assist students in visualising the extent or geographical-ness of the problem. For example, "teaching students with the notion of technologically supported learning, critical thinking, questioning and problem-solving skills are one of the main principles of science and technology teaching" (Balım, 2013, p. 338). The second category advocates for technology not as a supplement but as a replacement to teach physical or traditional classroom geography. Technology such as virtual reality or augmented reality are examples given to substitute outdoor learning (McClaren, 2019). While digital maps are also suggested to "be re-orientated to represent behavioural reality" (Cecioni, 2005, p. 288, Cardullo & Kitchin, 2019).

Students are able to deepen cognitive development and social learning within the comforts of remote learning. However, limitations are also found with technology being a full-on replacement, such that communication tends to be compromised. "ICT tools could create major differences in students' language skills. This made it difficult for teachers to keep up with communication levels between students in the classroom" (Sund & Gericke, 2020, p. 788). The last category highlights the use of technology as a critical 21st-century skill within pedagogy. These articles talk about digital technologies in general and instead focus on the importance of using technology in pedagogy due to advancements in society (Doering & Veletsianos, 2008; O'Flaherty & Liddy, 2018). These articles expound on the importance of information and communications technology for the rapid and continual advancements of technology in society. In this sense, the last category of technology also ties into the attitudinal aspects of learning.

The criticality of the hands-on approach can similarly come from framing it around learning by doing. These approaches include strategies that include transformative action, assemblages and social labelling. For example, one of the articles highlights how transformative geography enables students to "develop and undertake action based on a decision after recognising and understanding the issue" (Kirman, 2003, p. 93; Petersen & Barnes, 2020). This elaborates on how hands-on approaches can create a deeper understanding of the issue at hand, enabling potential dispositional changes including being more aware of geographical concerns.

Some pedagogies that deal with the development of attitudes and beliefs include storytelling and delivering information in an organic form. Some examples include "acknowledging cultural and contextual realities and revealing multiple perspectives" (Smithwick et al., 2018, p. 41). Students' interactivity and learning are enhanced, where interactive videos provide a portal for engaging students in complex systems thinking around the topic of sustainability. The examples where students produced videos for the purposes of assessment shows how these can be powerful learning opportunities.

The "process of storyboarding was an important reflective process" that led them to develop a genuine interest in the science topic (Gold et al., 2015, pp. 242–243). Through this approach, empathy is formed by placing oneself in a situation before being able to form new attitudes or rather new behavioural outcomes through empowering emotional learning in students.

There are other pedagogies that intertwine the knowledge, skills and attitude outcomes. Some of these include global and future citizenship. Pedagogies that build on global and future citizenships have a goal for students to be equipped with 21st-century skills and in advancing global competencies. They usually include a multi-dimensional view to function in a global society and being able to move productively into the future, such as introducing controversial issues for the analysis of different viewpoints, values and perspectives. One article states how global citizenship education prepares students to function in a global society, where "normative concepts may attain greater global influence by fostering discourse connections and broadening range of identified actors that actively promote them" (Kolleck & Yemini, 2020, p. 322). Through this, students are able to connect to the ethical, relational and political dimensions, shaping them to be informed responsible and active environmental citizens. Students are empowered to take action for a better community and their futures.

Pedagogies such as the inquiry approach encourage not only learning of knowledge and skills but also critical thinking. These approaches deal with cognitive processes and overlap with acquiring transformative behavioural or dispositional outcomes that signifies improvements in skills and changes in environmental behaviour or dispositional values and characteristics through places or a project. All of them essentially deal with having a problem identified. However, some of the methods highlight how they intend for students to solve a problem while others will elaborate on outcomes where students are able to develop their own solutions and have evidential behavioural or dispositional changes, for example, using issues through fieldwork that "actively involve students as a critical knowledge producer that create opportunities for them to reconstruct knowledge through experiential practice" (Elwood, 2004, p. 62; Luber, Fögele, and Mehren, 2020). Behavioural transformation can be seen in an article where "students can be taught that social responsibility as environmental actors go beyond decisions about their roles and political stances as members of local and global communities" (Steinberg, 1997, p. 117; Lawson, 2007).

This is where pedagogies at the nexus of attitudinal and behavioural learning outcomes include critical thinking approaches that focus on developing critical thinking in classroom settings that are not necessarily hands-on per se. Emphasis on critical thinking under these approaches does not necessarily translate to actionable results but rather, translation from basic skills such as absorbing facts to critically analysing and evaluating the information they have learnt. For example, one article calls for the expansion of active learning and critical pedagogy strategies to engage and empower students by looking at socio-environmental issues or from an environmental cultural lens (Petersen & Barnes, 2020, p. 7). Deeper learning here is achieved to evoke a sense of judgement and responsibility.

At the end of it all, to perhaps tie them all together will be the teacher-discourse approach. These approaches emphasise teacher's instructional methods adopted to improve student learning and planning resources. A wide array of strategies is mentioned, such as the use of technology in conceptual and transformational methods to aid, or rather act as an instrument to facilitate teacher instruction (Saribas, Kucuk, & Ertepinar, 2017; Sinha et al., 2017; Schultz & DeMers, 2020). They mostly act as a form of instructional intervention, where instructional methods such as demonstrations, micro-teachings and outdoor activities have an evidently positive effect on participant feelings. Ultimately, the outcomes presented target dispositional values, cognitive developments and transformative changes which target all areas of learning to do, to be, to know and live together.

Focusing on correcting misconceptions

Why is there an intentional focus on correcting misconceptions given the range of good pedagogies presented in the literature? Chapter 1 of this book began with a discussion of how climate change, one of the most urgent and current socio-scientific challenges of our day, has been shrouded in controversy, at least in the way the topic has been depicted in the media. In fact, the debate over climate change has developed over the last few decades. From denying that climate change is occurring, to admitting that while it is occurring, it may not be caused by people, to admitting that even if we are the source of the changes, the magnitude of change may not be as severe as has been reported. The recent trend, in which sceptics now believe that climate change is very certainly driven by human activity, but ask "so what?" and cite benefits such as greater food production in some regions of the world, is concerning. In spite of its prominence in political, economic and social discourses, there is still insufficient public interest in climate change mitigation that has resulted in widespread, collective and long-term action. People's reluctance to change their current lifestyles has played a significant role in this perceived disengagement. Three has been discussion on how students may not take on pro-climate behaviour and perhaps the lack of knowledge is a key contributing factor. Perhaps what is required is for individuals is to find meaning and value in the knowledge that is offered to them (Wi & Chang, 2019) and not just having knowledge. Factual knowledge that is provided to learners is insufficient to give them the meaning or reason to change their habits. For example, someone may be taught to raise the thermostat on an air conditioner but may not comprehend why this is necessary and in what ways this will contribute to mitigating climate change. We need to provide learners with the tools to turn the information they encounter into knowledge, and then take action based on that understanding. One of the key challenges in finding out what students know or do not know about the climate change topic is the persistent misconceptions that do not seem to be easily corrected.

Several studies reporting classroom interventions to address climate change misconceptions found that misconceptions persisted. For example, Gautier, Deutsch, and Rebich's (2006) simulated summit on global climate change with university undergraduate students indicated that, while the intervention increased students' awareness of the greenhouse effect, the bulk of the misconceptions identified in the pre-test persisted. They discovered that students' lack of metaconceptual awareness caused them to miss errors in their mental models even after being instructed about the issue. In another study, Lee, Lester, Ma, Lambert, and Jean-Baptiste (2007) utilised writing prompts to test fifth-grade students' awareness of the greenhouse effect and global warming. In a study unit that lasted over a school year, lectures, experiments and audio-visuals were employed as pedagogical tools. Similar to Gautier, Deutsch, and Rebich (2006), the authors found that while students demonstrated a more robust comprehension post-instruction, this did not transfer into accurate explanations of an enhanced greenhouse effect. In reality, the conflation of an enhanced greenhouse effect with ozone layer loss has persisted. The authors of both research studies concluded that enhanced comprehension does not imply the correction of misconceptions. If such misconceptions continue to exist, then we must find effective ways for teachers to help students correct them.

How do we correct misconceptions?

Misconceptions, according to the traditional notion of conceptual change, are alternative frameworks that must be replaced with accepted scientific perspectives (Posner, Strike, Hewson, & Gertzog, 1982). For a long time, the idea of misconceptions as barriers to effective learning has been featured in research and instructional practices, mostly in science education but also in other disciplines (Vosniadou & Verschaffel, 2004). The biggest criticism of the method came from constructivism proponents, who argued that rather than portraying errors as mistakes, pupils' prior conceptions are resources for cognitive growth (Smith, di Sessa, & Roschelle, 1994; Larkin, 2012). Smith, di Sessa, and Roschelle (1994) were particularly critical of instructional techniques that sought to replace or confront misconceptions, instead advocating knowledge refinement and reorganisation. Furthermore, educational techniques that encourage a slow and gradual process of re-crafting knowledge in recognition of students' lack of metaconceptual awareness of the structure of their prior knowledge have replaced the gestalt-type of knowledge reconstruction (Vosniadou, 2003).

Several studies have highlighted accurate knowledge about the causes of climate change as a major predictor for informed action and the creation of an individual's personal commitment to counteract the phenomenon's impacts. Indeed, false knowledge has no effect on actual behaviour modification in the same way that accurate knowledge does (Bord, O'Connor, &

Fisher, 2000). It seems to be odd that popular fallacies about climate change continue despite an abundance of knowledge about it from a variety of sources, from print to social media. It is indeed worth remembering that students do not approach a classroom as empty vessels waiting to be filled with knowledge. Indeed, individuals may have prior or pre-instructional knowledge about the material to be learnt, which is frequently flawed. These deeply ingrained and established assumptions have been shown to remain even after a learning session (Duit & Treagust, 2003). When there aren't enough resources to make sense of a situation, a person's natural instinct is to gravitate toward what appears to be the most reasonable explanations. One approach is to seek facts and opinions from those deemed to have intellectual or moral authority on the subject (Bulkeley, 2000). What results is a tightly woven set of notions made up of fragments of facts, fiction, opinions and myths. This tapestry of personal hypotheses becomes the common-sense reasoning to describe an event over time (Schumacher et al. in Duit, 1999).

The literature indicates that across localities, general public perception about climate change and related concepts are riddled with various alternative interpretations and beliefs, often leading to unorthodox and unscientific explanations of the issue. The greenhouse effect, for example, is often confused with other environmental problems (Bord, O'Connor, & Fisher, 2000; Pruneau et al., 2001). Henry (2000), in studying the public perceptions of laypeople about global warming, noted three prevailing misconceptions. First, that ozone depletion causes global warming. Second, the effects of global warming will be sudden and catastrophic. Third, anthropogenic contributions to climate change are not significant. Henry also noted in his fieldwork observations that it is rare for participants to not "incorporate the famous hole in the ozone layer with the entirely independent problem of global warming" (2000, p. 27).

This widespread public confusion on the science of global warming is well noted by McBean and Hengeveld (2000). This could be due to ineffective communication skills of scientists, compelling arguments of contrarians and the media's failure to cover relevant events with an eye on scientific debate about detail and significance.

Duit (1999) contends that learning may not be as simple as it looks, as teachers themselves may "hold conceptions of science concepts and processes that are not in accordance with the science view and often are similar to students" pre-instructional conceptions' (in Duit & Treagust, 2003, p. 298). In the case of climate change, several studies highlight the need for teachers to upgrade, if not unlearn, their content knowledge (Fortner, 2001; Padpadimitriou, 2004; Hestness, McGinnis, Riedinger, & Marbach-Ad, 2011). Thus, Hewson et al. in Treagust and Duit (2008) strongly suggest going back to restructuring educators' misconceptions as a first step.

According to research, students are largely perplexed about climate change, frequently conflating related notions with another environmental issue, the most common of which is ozone depletion. This is true across age groups and years in various research locations (Francis, Boyes, Qualter, and

Stanisstreet, 1993; Koulaidis & Christidou, 1999; Lee, Lester, Ma, Lambert, & Jean-Baptiste, 2007 among others). Furthermore, they had difficulty distinguishing between notions such as weather and climate (Gowda, Fox, & Magelky, 1997; Padpadimitriou, 2004). Patterns demonstrate that they frequently mix up scientific and conceptual aspects from the two atmospheric events (Kilinc, Stanisstreet, & Boyes, 2008). Other misconceptions identified are the following:

1. Inflated estimates of temperature change (Gowda, Fox, & Magelky, 1997)
2. Confusion between the CFCs (Gowda, Fox, & Magelky, 1997)
3. Global warming is easily discernible in the short term (Gowda, Fox, & Magelky, 1997)
4. All environmental problems cause climate change (Boyes, Boyes, Chuckran, & Stanisstreet, 1993; Francis, Boyes, Qualter, & Stanisstreet, 1993; Gowda, Fox, & Magelky, 1997; Padpadimitriou, 2004) (e.g., lead pollution causes greenhouse effect; radioactive contamination and acid precipitation are symptoms of the greenhouse effect)
5. Confusing weather and climate (Gowda, Fox, & Magelky, 1997; Padpadimitriou, 2004)
6. Understanding of the fundamental societal consequences of drastic reduction of CO_2 emission (Andersson & Wallin 2000)
7. Carbon dioxide destroys the ozone layer (Rye, Rubba, & Wiesenmayer, 2007); this is related to the confusion between the greenhouse effect and ozone layer depletion
8. All acts appropriate to care for the environment could solve the problem of global warming (Boyes & Stanisstreet, 1993; Fortner, 2001; Daniel, Stanisstret, & Boyes, 2007; Punter, Ochando-Pardo, & Gracia, 2011)
9. Greenhouse effect causes skin cancer (Jeffries, Stanisstreet, & Boyes, 2001)

Replacing long-held schema is a familiar challenge. For example, Cordero, Todd, and Abellera's (2008) 15-week course on weather and climate failed to change university students' misconceptions. Further, Lee, Lester, Ma, Lambert, and Jean-Baptiste's (2007) experimental study involving fifth-grade students across the United States discovered that students continue having difficulty differentiating between the greenhouse effect and the enhanced greenhouse effect, as well as ozone layer depletion as a different concept, even after discussion of such topics for post-assessment. It is believed that the persistence of misconceptions is due to erroneous learning they have been exposed to in their foundational year courses that is further exacerbated by media reports (Cordero, Todd, & Abellera, 2008).

To correct misconceptions, McNeill and Vaughn (2012) posit that learning should equip students to successfully connect the dots from knowledge to actions; their conceptual understanding of the climate change phenomenon must be deep enough to guide them into active involvement. The pedagogy must be purposefully designed to target misconceptions and in creating an environment for students to think critically. Collaborative learning (Mason & Santi, 1998) is seen as one pragmatic approach to deconstructing and restructuring prior or pre-instructional knowledge.

Central to conceptual change approach is the restructuring of prior understanding (Duit & Treagust, 2003), facilitating the rejection of erroneous schema and building up a new, accurate knowledge base (Atasoy, Akkus, & Kadayifci, 2009). It is a complex process that takes time to implement and, as documented in not a few cases, may not always be successful (Limón, 2001). Rejection is most noted when prior or pre-instructional knowledge is deeply rooted in the daily life experiences of learners who may view the newly introduced conception as unintelligible and not plausible (Duit, 1999). In the climate change discourse, for example, students find it most conceivable that the hole in the ozone layer lets in more sunrays, thus, an increase in global temperature. They then confuse chlorofluorocarbons as similar to greenhouse gases and lump the two environmental phenomena as one and the same.

One way to restructure prior knowledge is to create conflict in a student's pattern of thinking. By presenting an idea that does not appeal to common sense or may even be radically different to what is believed as truth, a learner is encouraged to resolve the predicament (Stepans, 2003). The role of the educator is to skilfully guide the unravelling of new conceptions.

For conceptual change to happen, four conditions must be met. First, the learner must be *dissatisfied* with a current conception. Irregularities in the explaining power of an existing frame of thought must become too obvious to be ignored. Second, the new concept must be *intelligible* to the learner. One must be able to fully get the picture of how the elements inherent in a concept can explain a phenomenon. Third, it must be *plausible*. The new concept should aptly address the dissatisfaction felt by the learner with the old concept. In short, it must be able to solve the problems that the old concept was not able to. Fourth, the new concept must be *fruitful*. It must have the capacity to prompt the learner to explore new areas of learning (Posner, Strike, Hewson, & Gertzog, 1982).

"Closing the Hole in the Sky" is a study on a pedagogical approach to rectifying students' misconceptions in CCE through refutation texts that specifically highlighted misconception (bogus text) and the correct text (Chang, Pascua, & Ess, 2018). This study is based on instruction-induced conceptual change, a sort of conceptual change that involves systematic instruction via an intentional top-down, intentional learning mechanism (Vosniadou, 2007). In looking at conceptual change, this method supports Hatano and Inagaki's (1991) and Inagaki and Hatano's (2003) viewpoint of merging individual and constructivist approaches on a sociocultural framework. It used the approach of inducing conceptual change via a comprehension activity conducted by a teacher and supported by peers. This is done by giving a multiple-choice problem with alternative answers that represent typical misconceptions held by learners.

The instruction was also altered to a pedagogically led approach in order to explore if the teacher's pedagogy may potentially play a role in reinforcing mental models and conceptual understanding in the learners. This was achieved through the use of a lesson study (LS) method, in which the teacher delivering the intervention collaborated closely with other teachers and researchers to design instructional tactics (Chang, Pascua, & Ess, 2018). To determine if the intervention was effective, a content analysis of homework submitted by students was performed using the Climate Change Diagnostic Test (CCDT) (Chang, Pascua, & Ess, 2018).

In a pre-test and post-test performed before and after teaching, the CCDT was utilised as a quantifier for misconceptions. Students were allowed to become acquainted with the text material in order to make it easier to use. Considerable differences were detected between pre-test and post-test scores after instruction, with post-test scores demonstrating significant gains in understanding after the activity (Chang, Pascua, & Ess, 2018). The findings indicated that a deliberate teaching approach supplemented with refutation-based instruction had a positive influence on students' general understanding of climate-related concepts. Instead of teaching the issue of climate change in general, the primary goals were to decouple the notion of particular misconceptions (Chang, Pascua, & Ess, 2018). These include dispelling the misconception that natural greenhouse effect is the same as the enhanced greenhouse effect and explaining how ozone depletion exacerbates the enhancement of the greenhouse effect. Overall, the study demonstrated that pedagogically led instruction in the form of direct refutation played a pivotal role in student conceptual understanding. Strong statistical improvements in identifying and correcting misconceptions underlined the favourable impact of students' conceptual understanding and reasoning.

Refutation text and instruction

Given the empirical evidence that supports refutation instruction as a way to correct students' misconceptions about climate change topics, we endeavour to provide the details of the pedagogy for educators' reference. A refutation text is a text structure that directly challenges readers' misconceptions by strategically presenting the alternative conception, which is then contradicted with accurate facts. According to Guzzetti's (2000) meta-analysis of seventy articles on text format and usefulness, refutation text is the most useful type of text as rated by elementary, high school and adult learners. Furthermore, as compared to other educational approaches such as demonstrations, cooperative discussions and non-refutational expository writing, only refutation text shows long-term advantages.

Studies comparing the effects of refutation and non-refutation texts illustrate how readers' cognitive processes alter while engaging with the texts. Kendeou and van den Broek (2007) discovered that when new information in the text contradicts prior knowledge, readers adjust their cognitive processing. Further research (van den Broek & Kendeou 2008) revealed that the simultaneous activation of accurate and wrong concepts while reading plays an important role in knowledge updating, resulting in a significant rise in conflict awareness. Ariasi and Mason (2011) thus concluded that better learning is indicated by a reader focusing longer on key components of the text as facilitated by strategic reading. Metacognition is encouraged by refutation text (Tippett, 2010), such that readers who have misconceptions are more likely to identify discrepancies in their previous knowledge when faced with an alternative explanation.

Refutation texts encourage readers to compare their own views with new and unfamiliar facts, which Hynd (2001) considers to be a persuasive aspect of refutation texts. Because both sides of a topic are addressed, the text can be regarded as an expert source. Furthermore, refutation text is useful in finding support for scientific ideas, learning new language required to communicate understanding and learning new concepts (Guzzetti, Williams, Skeels, & Wu, 1997). Text based on refutation, on the other hand, may not always provide favourable outcomes. Students may choose to disregard the material entirely while adhering to their alternate conceptions. According to Guzzetti, Williams, Skeels, and Wu (1997), in such circumstances, refutation was either ineffective or students' reading methods were insufficient to induce conceptual change. As a result, text as a stand-alone element is insufficient to challenge or modify learners' misconceptions (Broughton, Sinatra, & Reynolds 2010). To guarantee long-term conceptual change, teachers should provide guidance and support through refutation instruction (Guzzetti, 2000).

Refutation instruction

The lesson design was developed collaboratively between the researcher and the teachers (Chang, Pascua, & Ess, 2018). The study team met with two teachers to define the research aims and the expected learning outcomes from the intervention, as well as the use of refutation texts and the pedagogical approach to teaching the topic. The intervention was divided into four different sessions, each of which addressed a different aspect of weather and climate: (1) the greenhouse effect, (2) the enhanced greenhouse effect, (3) climate change and (4) the impacts of climate change. There were two sets of refutation texts: (1) text on a sheet of blue paper (the blue paper), which contains alternative conceptions about climate change and (2) text on a sheet of yellow paper (the yellow paper), which directly refutes the misconceptions in the blue paper. For the first two sessions, the items in both papers were presented in numbered order for ease of use. For the third and fourth sessions, it was changed to a paragraph form. The students were instructed to work in pairs and freely discuss with their peers. The blue papers were given to the students first, with the instruction to highlight any words or phrases that they thought were incorrect. The students had ten minutes to finish the activity. Following that, the yellow papers were distributed. After the teacher read aloud the expert text, the students were instructed to annotate the blue paper, correcting the errors in the items using the expert text as a guide (Chang, Pascua, & Ess, 2018).

Natural greenhouse effect

The greenhouse effect session focused on how the phenomenon is a natural process that allows life on Earth to exist. The focus of the lesson was on debunking the myth that the natural greenhouse effect and the enhanced greenhouse effect are the same. It also highlighted the elements and processes involved in terrestrial heat absorption. "Too warm for life to thrive", "direct sunlight" as the source of heat that is absorbed, "shortwave radiation", the greenhouse effect as a "layer of gases", and "water vapor is not a greenhouse gas" were among the refutation text's (yellow) counterpoints to the bogus text's (blue) statements. Students were also given an illustration of the greenhouse effect to comment on.

The diagram received no annotation from ten students. The most common annotations were (1) the greenhouse gases are not in a layer, (2) the presence of cars and factories in the illustration refers to the enhanced, not the natural, greenhouse effect and (3) the diagram was updated to include shortwave and longwave radiation. The refutational text and instruction successfully restructured six misconceptions previously held by the students in the pre-test (Chang, Pascua, & Ess, 2018). The following are the misconceptions correctly refuted:

1. Water vapour is not a greenhouse gas.
2. Water vapour cools, not warms, the atmosphere.
3. The greenhouse effect occurs mostly in the ozone layer.
4. The heat released by the Earth gets absorbed mostly in the ozone layer.
5. More heat is absorbed by the atmosphere through the ozone layer.
6. Longwave radiation from the sun makes it possible for greenhouse gases to absorb heat.

Water vapour is considered a greenhouse gas after the refutation instruction. The majority of respondents in the pre-test believed that water vapour does not have the ability to absorb heat. The distinction between the greenhouse effect and ozone layer depletion is noteworthy. The coupling of these two atmospheric phenomena was the greatest hindrance to learning new concepts accurately, according to the literature and primary data gathered by the researchers. Surprisingly, this research indicates that they have been successfully untangled (Chang, Pascua, & Ess, 2018).

Enhanced greenhouse effect

The main goal was to disentangle how the hole in the sky or depleted ozone causes or exacerbates the greenhouse effect. CFCs were highlighted as a confounding variable in students' conceptual understanding. CFCs were deemed the key element that required extensive explanation because they are both a greenhouse gas and a gas that damages the ozone layer. During the discussion on natural climate change, one student brought up the issue of ozone depletion prematurely. The role of humans in enhancing the greenhouse effect was also addressed as a common misconception. This session included in-depth explanations from the teacher as well as detailed illustrations. Several students were also asked to copy the illustrations and explain them to the rest of the class (Chang, Pascua, & Ess, 2018).

For many of the items, the connection between the ozone hole and the increased greenhouse effect was either debunked or reduced in frequency. Given the extent of misconceptions discovered in the pre-test, this is an impressive result. The burning of fossil fuels, on the other hand, is thought to contribute to the greenhouse effect by heat absorbed by longwave or shortwave radiation from the sun, as well as the processes that contribute to heat, rather than the burning itself. Longwave radiation is also responsible for warming the Earth, according to the students (Chang, Pascua, & Ess, 2018).

Climate change

The third session focused on climate change as an atmospheric event. The impact of such atmospheric events on Earth's physical geography was discussed, as well as the patterns of cooling and warming, the events that led to such changes and the impact of such atmospheric events over Earth's geologic time. The students' main association with the climate change phenomenon is with global warming, according to the pre-test. The refutation text and instruction emphasised that the phenomenon does not only apply to global warming but also to global cooling as a general introduction to the topic. The natural changes in climate were also discussed, to counter the perception that they are primarily due to anthropogenic causes. There was a significant difference in the pre-test and post-test scores.

Even after the instruction, more students were inclined to believe that an increase in solar radiation is not responsible for the enhancement of the greenhouse effect, despite the fact that several misconceptions were debunked. The ozone layer confusion resurfaced, with some claiming that an increase in solar radiation has an impact on the ozone layer. In addition, more attention should also be accorded to the students' belief that daily weather patterns are indicators of climate change. However, the majority of the annotations on the blue text highlighted the texts without providing corrections, while others did not highlight the relevant phrases at all. This indicates that the transition from numbered to paragraph format in the texts discouraged students from effectively annotating the materials (Chang, Pascua, & Ess, 2018). This is certainly something to take note of and a suggestion will be to adopt the use of numbered paragraphs/items throughout the refutation text to ease the cognitive load off the reading.

Effects of climate change

Class sessions on the impact of climate change focused on correcting false associations between the phenomenon and other environmental issues like earthquakes, tsunamis, acid rain and skin cancer. The causes of rising sea levels were also highlighted.

Misconceptions about sea level rise have been thoroughly debunked. Students were also able to distinguish between atmospheric and tectonic plate movements. The conception of the likelihood of an earthquake or tsunami occurring as a result of climate change has decreased. Acid rain and skin cancer are both examples of this too. In this session, students annotated less than in the previous two. It was suggested in the third session on climate change that the change to paragraph format for the refutation text discouraged students from properly annotating the blue paper (Chang, Pascua, & Ess, 2018). The students' decreased engagement with the material was attributed to the text's reduced readability and ease of reading (Meyer, 2003).

Which pedagogy should I choose?

Does a refutation-based pedagogical tool help secondary school students understand climate change? There is strong statistical evidence that the use of refutation-based instruction resulted in a significant improvement in addressing misconceptions, implying that it had a positive effect on students' overall understanding. Tackling and teaching against misconceptions does improve students' acquisition of key concepts, according to Muller, Bewes, Sharma, and Reimann (2008). Furthermore, taking a sociocultural perspective on conceptual change aided in a close examination of what works and what doesn't, involving teachers and students in a close knowledge restructuring process. Nonetheless, Chang, Pascua, and Ess (2018) did not claim that all misconceptions had been corrected. In fact, there were a few misconceptions that persisted. One example is the conflict between infrared, ultraviolet and visible light as types of absorbed and re-emitted radiation. Nonetheless, any attempt at refutation, or even conceptual change, for CCE should not be viewed as merely re-inventing the wheel. It should equip children with a correct, coherent and comprehensive understanding of climate change concepts so that they can engage with information about the topic in order to protect our shared environmental future.

But refutation instruction is not the only pedagogy presented in this chapter. Pedagogies that help correct misconceptions were discussed in a meta-analysis of the works published in key education research journals in environmental and geographical education. The pedagogies reviewed included hands-on approaches, conceptual approaches, use of online & technological tools and pedagogies for disciplinary approach, among others. A hands-on approach might result from structuring it around learning by doing. Transformative action, assemblages and social labelling are among the tactics used in these approaches. One of the papers, for example, emphasises how transformational geography helps students to create and take climate action based on their comprehension of the concerns presented.

Students can benefit from conceptual approaches in developing analytical frameworks for thinking about content knowledge. Skills such as reading comprehension or critical analysis may also be included in conceptual approaches. In climate science, we may also employ ideas like adaptation and resilience to plan or arrange a curriculum that reflects the students' learning progressions. For example, students may begin by studying how climate science is organised, then go on to understand what adaptation and resilience ideas entail, followed by a lesson on climate change response. Other topics that might be taught and revisited during the year include how human activity impacts the Earth's ecosystem and climate.

There are other pedagogies that promote skill development too. This includes online and technological approaches that focus on technological applications and the integration of digitalisation into student learning. When technology is employed as a learning aid, technology can be used to teach physical or conventional classroom topics as a substitute rather than as a complement. Technological approaches can also focus on supporting the attitudinal aspects of learning. Disciplinary approaches, on the other hand, focus on a single discipline or holistic approaches such as multi-, inter- or cross-disciplinary approaches to increase comprehension and cognitive growth. Articles focusing on a single discipline primarily advocate for CCE to be taught within that discipline. This opens up possibilities for interdisciplinary approaches as well.

Summary

In truth, there isn't any single pedagogy that is better than the others. But the chapter focused on refutational instruction to correct student misconceptions in hopes that it will make a difference in how students will take the knowledge learnt beyond and make behavioural changes for climate change. Indeed, the refutation instruction approach can be used in combination with other pedagogical approaches such as collaborative learning and inquiry-based learning, for example.

Nonetheless, studies in the literature support the claim that global climate change literacy as a goal meets a tough adversary in the shape of persistent alternative conceptions. Indeed, the onus of preparing a generation of critical thinkers about climate change rests not only on a "storm-ready" curriculum, but also on teachers' pedagogical readiness and awareness of the complexities involved with prior knowledge that cloud students' understanding of the topic (Kagawa & Selby, 2012). Next, we shall turn our attention to teachers' readiness in CCE.

6 Preparing teachers for CCE

From studies around the world, a vicious cycle is observed of students not being given enough and accurate education on climate change, of teachers not knowing how or what to teach and a public that is misinformed of these. The literature is indicative of the lack in substance in the understanding of climate change and the prevalence of misconceptions about the phenomenon issues (Boon, 2010; Hestness et al., 2011; Chang & Pascua 2016). Further, the texts reviewed explicate that awareness and self-attested knowledge of climate change issues do not necessarily equate with understanding. Knowledge, in this sense, could take the form of an accurate understanding or spurious interpretation of information. Both those with accurate and spurious knowledge may believe that climate change is occurring, for example, and they may even have the same intention to act on such belief. In terms of behaviour change, however, spurious knowledge tends to not translate into action and those with accurate understanding are more likely to act towards addressing the issue. Positive attitude is by itself not a sufficient alternative to accurate knowledge. A pro-environmental stance without accurate understanding eventually may only lead to an attitude of apathy and nonchalance. Misconceptions, if uncorrected and held on to for an extended period of time, cannot be easily dislodged. Left unchecked, these beliefs are likely to be carried into adulthood, further feeding into the knowledge deficiency-inaction loop.

Disentangling flawed models of understanding is a complicated and lengthy process. Going back to Bord et al.'s (2000) assertion that accurate knowledge precedes sensible decision-making and purposeful action, it is therefore paramount that knowledge transfer be done in a systematic and thoughtful manner right where learning is the only business that matters – in school classrooms. Education must endeavour to help the student comprehend the complexity of climate science, enable the learner to have an open mind and critical approach to possibilities and limitations as they present themselves and to be challenging enough to encourage them to argue and decide on socio-scientific issues. Indeed, Bangay & Blum (2010) maintain that education's potential in contributing to adaptation and mitigation measures especially in penetrating mainstream development thinking is

DOI: 10.4324/9781003093800-6

largely untapped. The trends of policy discussions have been, by far, focused on technical solutions without seriously engaging with the issue of teaching climate change education (CCE). Thus, it is pertinent to first determine what teachers know in order to advance the issue.

Teachers are the key stakeholders in the climate change discourse. One of the common problems associated with learning climate change is that students often have misconceptions with regard to concepts such as the greenhouse effect (GHE). The greenhouse effect refers to the trapping of heat by greenhouse gases like carbon dioxide in the atmosphere, resulting in the Earth warming up. It is often wrongly assumed, however, that the hole in the ozone layer caused by chlorofluorocarbons, allows more sun rays to enter the Earth and thus, warming it up (Punter, Ochando-Pardo, & Gracia, 2011; Kilinc, Stanisstreet, & Boyes, 2008). Misconceptions of this kind may hinder students' ability to take the correct step to tackle climate change. Thus, teachers play a role in addressing such misconceptions. This begs one question: Are teachers equipped with the relevant knowledge and skills to help students correct these misconceptions?

What do teachers know about climate change?

In studies conducted to assess teacher trainees' knowledge of climate change, it was found that future classroom teachers themselves hold the same erroneous understanding as that of secondary students (Ocal et al., 2011; Khalid, 2003; Ikonomidis et al., 2012). This is an area of concern as students typically trust their teachers' knowledge authority on the topics they teach (Gowda et al., 1997). If teachers themselves have misconceptions, then they are more likely to pass on these same erroneous understandings to their students. Hence, education scholars recommend strategies to correct such misconceptions. These include employing student-centred methodologies when teaching teacher trainees. For example, teacher trainees can carry out in-depth research on a particular topic and present and discuss their findings with their peers. Through such discussion, they will be exposed to a wealth of perspectives and such exchange of information will allow them to evaluate and correct their pre-existing conceptions (Khalid, 2003).

The literature generally points to a dearth of knowledge among educators on climate change. Fortner (2001) of Ohio, USA, points out that while teachers do have an interest in teaching about climate change, they nonetheless have inadequate knowledge to be fully competent for the task. Like their students, some teachers are unable to distinguish among related environmental issues. They have the tendency to lump all environmental problems together with climate change concepts. Basic knowledge about the phenomenon is lacking, and characteristics, causes and effects of environmental problems get lumped together with concepts on climate change. Foster believes that teachers' inadequate knowledge and lack of confidence in teaching topics on climate change could be traced to their lack of background information in their university studies. They are then likely to possess deeply held misconceptions and confused issues, possibly passing them on to their students. Barriers to meaningful understanding of the issue are further exacerbated by an overcrowded curriculum, lack of interdisciplinary training as well as the reluctance of teachers to discuss a topic that is inherently intangible and not immediately observable (Fortner, 2001). Fortner (2001) further explicated that if this trend continues, global climate change information may be left to television shows, so that misconceptions may go unchallenged and no education at all may occur.

In a research study involving student teachers in Greece, it was found that prospective primary school educators hold misconceptions and misunderstandings concerning climate change, they confuse weather with climate and incorrectly relate climate change with issues such as acid rain and environmental pollution (Padpadimitriou, 2004). In the same light, Australian pre-teachers in Boon's (2010) study showed that future educators hold these same erroneous beliefs. This study pointed out the need to develop tertiary courses to improve pre-service teachers' knowledge and address gaps in curricular topics.

Hestness et al. (2011) elaborate that confusions and misconceptions could be addressed by merely clarifying the issues with the prospective teachers. The research discussed how trainee teachers were subjected to a module on global climate change. They were assessed on their knowledge, pedagogy, awareness of resources to address challenges with learners and ability to engage in meaningful conversation. The authors state that through the module, positive impacts were observed with regard to correcting erroneous knowledge.

In all, the literature sums up that teachers, students and the general public hold an incorrect understanding of climate change. It is a complicated cycle that starts with educators' lack of understanding of the phenomenon that is then passed on to learners and further exacerbated by incomplete and inadequate information gathered from alternative sources. The solution to this quandary may well be found in the classroom.

The need for effective teaching strategies in bringing the issue of climate change into the classroom cannot be overemphasised (Boyes & Stanisstreet, 1993). CCE should empower the learner to realise that he/she has a stake in how the future will be defined (Schreiner, Henriksen, & Hansen, 2005). Svihla & Linn (2012) support this idea of a redesigned curriculum, further stressing the importance of using evidence from student work to guide iterative refinement of instruction. Boyes, Chuckran & Stanisstreet (1993) also encourage educators to work at disentangling various issues and sources of confusion but cautioned against the tendency of espousing a non-holistic approach. In addition, Lee et al. (2007) put forward the idea of incorporating students' alternative conceptions and pockets of learning difficulties into curriculum materials and the professional development of teachers. Further, Gowda et al. (1997) suggest collaborative activities among scientists and educators in coming up with innovative curricula using fine-tuned concepts, up-to-date design and materials.

A study concerning Chemistry teachers in Germany was examined by Feierabend, Jokmin & Eilks (2011) where they found that the teachers felt that it was important to teach about climate change in the classroom. However, they mentioned that there is no general agreement about the teaching of climate change in the Chemistry curriculum and the amount of time allocated to spend teaching about it. Wise (2010) conducted a survey on science teachers on the teaching of climate change. Teachers surveyed are supportive of teaching climate change. Most teachers taught about the scientific aspect of climate change and also discuss the controversy surrounding it. It was found that most Earth Science teachers include climate change in their formal teaching while most teachers of other science subjects taught climate change informally. Teachers were motivated to teach climate change if the topic is included in the curriculum; they perceived it is important for students to understand about it; and they themselves have a good understanding of the topic. Misconceptions about climate change were also discovered among teachers. Based on the findings, the author recommended developing professional development that provides teachers with the knowledge to overcome their misconceptions as well as discussing the controversy of climate change in the classroom. Furthermore, professional development should also be directed at teachers across different subjects.

Another more elaborate study, focusing on current practices in CCE was carried out by the National Earth Science Teachers Association (NESTA) in the United States (Johnson, 2011). Firstly, it was found that teachers expressed that they were slightly or extremely comfortable teaching about climate change. Secondly, with regard to challenges, teachers faced difficulties in addressing students' misconceptions and the notion that climate change is not a result of human activities or not happening to students, parents, administrators or other community members. Thirdly, the majority of the respondents noted an increase in positive attitudes when teaching climate change. Fourthly, teachers rely on resources offered by government agencies as well as programs that have a long track record. And finally, teachers would prefer learning more about climate change at in-persons workshops offered at a location near them.

Following such studies, NESTA recommends developing professional development courses to prepare teachers in a few areas. Firstly, teachers can be provided with the knowledge to ensure that they have updated knowledge. Next, effective teaching strategies can be taught to teachers to allow them to better engage their students in the study of climate change and how students can deal with the issue of climate change in a holistic approach. Finally, there is also a need for professional development for teachers on the effective use of resources to support the teaching and learning of students (Johnson, 2011). While these studies generally point to a need to provide professional development support given the multiple issues teachers face in carrying out CCE, the contexts for these classrooms differ from those of the context within which most of the key examples used in this book draw from. Hence, it is essential to examine the case of Singapore.

Teacher readiness in Singapore

The curriculum is not just a planned document, but it has to be implemented and enacted in classrooms. Hence, how it is implemented and enacted depends on the key stakeholder – the teacher. Rather than determining what teachers know or do not know in the Singapore context, a study by Chang & Pascua (2018) was conducted to examine how teachers perceive of their own readiness to teach about climate change. A content analysis of Ministry of Education syllabi was first undertaken to ascertain the coverage of climate change topics in the formal curriculum. Consequently, an online ten-item questionnaire was developed and sent to teachers as a dipstick study in understanding how the formal curriculum is implemented and enacted in classrooms, from the results of the content analysis. Of the 55 Secondary Geography teachers invited, 29 participated in the online survey. Of these, 15 teachers are between 21 and 30 years of age, the remainder 9, 2 and 3 are aged 31–40, 41–50 and above 50, respectively. Table 6.1 shows the distribution of the teachers by the years of teaching experience.

The teachers were asked to rank their perceptions for four domains, namely:

1. Curriculum, curriculum design and implementation
2. Instruction and classroom teaching and learning
3. Outcomes of CCE – students' understanding; and
4. General perceptions

Implementing a CCE curriculum

In terms of curriculum design and implementation, the teachers felt that CCE is present in the national as well as the school curriculum (72% for both cases). In fact, they agree (76%) that CCE is featured in textbooks. However, teachers felt that support from resources (65%) and professional development courses and platforms (75%) to discuss issues on CCE are lacking. In terms of assessment, the perception was that while CCE is evaluated in formative class assessment (41%), it tends to be missing from summative assessment (83%). Table 6.2 lists the detailed responses on curriculum issues.

Table 6.1 Years of teaching experience of survey respondents

Number of years of teaching	*Frequency*
<2	12
2–5	3
6–10	5
11–15	5
>15	4

(Source: Chang, 2014 p. 83).

Table 6.2 Survey items about curriculum, curriculum design and implementation

Items	*SD (%)*	*D (%)*	*N (%)*	*A (%)*	*SA (%)*	*Mean score*
CCE[a] is explicitly taught in the national curriculum	0	14	14	**62**	10	3.68
CCE is explicitly taught in my school's curriculum	0	10	17	**48**	24	3.83
CCE is infused into existing curriculum	0	10	10	**59**	21	3.91
Topics on CCE are included in school textbooks	0	10	14	**55**	21	3.87
There are sufficient resources in textbooks, workbooks, teacher's guides and CD-ROMS for CCE	0	**41**	24	34	0	2.90
There are professional development courses to support me in CCE	3	**41**	31	21	3	2.77
There are professional platforms to support me in discussing CCE with fellow teachers	3	**55**	21	21	0	2.60
I feel that I need more support in designing lessons on CCE	0	10	17	**59**	14	3.77
Students are assessed at the national level exams for their understanding of climate change issues	0	31	**52**	17	0	2.86
Climate change issues are included in school summative assessments	0	24	34	**41**	0	3.14
Climate change issues are included in classroom formative and continual assessments	0	24	24	**48**	3	3.27

(Source: Chang, 2014 p. 84).

a Teaching and learning of issues on the causes, impact, mitigation and adaption of climate change will be referred to as Climate Change Education in this survey, abbreviated as CCE.

Classroom instruction for CCE

In terms of the actual teaching and learning of CCE in the classroom, 59% of the teachers perceived that they are confident in teaching CCE well. 90%, 100%, 96% and 100% of the teachers also felt that it is important to teach knowledge, concepts, skills and attitudes about CCE, respectively. Indeed, teachers strongly agree that the behavioural changes and attitudes are the

Table 6.3 Survey items about instruction and classroom teaching and learning in CCE

Items	*SD (%)*	*D (%)*	*N (%)*	*A (%)*	*SA (%)*	*Mean score*
I feel confident in designing good CCE lessons	0	14	28	**52**	7	3.55
It is important to teach the facts for CCE well. For example, average global temperatures have increased by 0.8°C over the last 160 years	0	0	10	**59**	31	4.21
It is important to teach the concepts of CCE well. For example, there is a difference between the natural and enhanced GHE	0	0	0	**59**	41	4.41
It is important to teach the skills for CCE well, for example, the ability to interpret the graphs and tables in the IPCC report	0	3	0	**48**	**48**	4.38
It is important to teach the attitudes and values for CCE well. For example, behavioural changes in students show that they have understood the need to mitigate climate change is like switching off the lights while leaving the room	0	0	0	38%	**62**	4.62
It is important for students to be able to discern and critique reports on climate change	0	0	0	**62**	38	4.38
It is important for students to raise questions about issues of climate change	0	0	3	45	**52**	4.49

(Source: Chang, 2014 p. 86).

most important aspects of CCE. 100% and 98% of the teachers felt that is important for students to be able to discern and critique reports on climate change and to raise questions about issues in climate change, respectively. Table 6.3 lists the detailed responses about the teaching and learning process in the classroom.

Perception of students' learning

However, the learning outcomes that teachers have noticed show that while students learn CCE facts well, there is little evidence from teachers' perceptions that they have learnt the skills and concepts of CCE well. In fact, teachers disagree that students are able to critique reports on climate change. However, teachers feel that students are able to raise questions about issues

Table 6.4 Survey items about students and their understanding of climate change

Items	*SD (%)*	*D (%)*	*N (%)*	*A (%)*	*SA (%)*	*Mean score*
My students generally know the facts for CCE well. For example, average global temperatures have increased by 0.8°C over the last 160 years	0	**34**	24	**41**	0	3.04
My students understand the concepts of CCE well. For example, there is a difference between the natural and enhanced GHE	0	**31**	**31**	**34**	3	3.06
My student have the skills for CCE well, for example, the ability to interpret the graphs and tables in the IPCC report	0	**41**	31	24	3	2.86
My students show that they understood the need to mitigate climate change like switching off the lights while leaving the room	0	7	28	**66**	0	3.62
My students are able to discern and critique reports on climate change	0	**45**	34	21	0	2.76
My students are able to raise questions about issues of climate change	3	17	34	**45**	0	3.19

(Source: Chang, 2014 p. 87).

on climate change and they have understood the need to mitigate climate change. Table 6.4 lists the detailed responses of the teachers' perception of student learning outcomes.

General impressions about CCE

Teachers have a generally positive attitude toward teaching CCE and believe that CCE will help their students become better global citizens. They also feel that it is important to make CCE more explicit in the curriculum. Table 6.5 shows the overall impression about teaching CCE.

While the existing curriculum clearly has topics that are directly related to CCE, in future revisions of the national syllabus, CCE can be made more

Table 6.5 General impressions about teaching CCE

Items	*SD (%)*	*D (%)*	*N (%)*	*A (%)*	*SA (%)*	*Mean score*
I hate teaching CCE[a]	21	**69**	3	3	3	3.99
I think it is important to teach CCE	0	3	3	**55**	38	4.25
I think CCE can be infused more effectively into our curriculum	3	0	7	**48**	41	4.21
I think CCE should be infused across all subjects in the curriculum	3	3	10	**52**	31	4.02
I think CCE should be made more explicit in some subjects in our curriculum	3	3	10	**48**	34	4.01
I believe teaching CCE will make my students better global citizens	3	0	7	41	**48**	4.28

(Source: Chang, 2014 p. 87).
a Note that this is a negatively phrased question. Mean score is calculated by reversing the scale.

explicit. One teacher opined, "I think more can be done to impart the impact of Climate Change on the Environment and what actions can the individuals take to reduce the impacts". Another teacher commented that CCE is "more explicitly taught in Sec 2 GE syllabus". At the upper sec levels, "CCE is taught as a sub-component in the topics of impacts of deforestation/development. It should deserve a greater weighting as discussions would be more in depth with the older students and in the move towards greater global citizenry". This is corroborated by another teacher that "currently, CCE is better reflected in lower secondary education, especially secondary 2s. It is only peripherally mentioned in elective geography through vegetation and it is a very small part". Indeed, the survey was conducted prior to the implementation of the current Geography syllabus in 2013.

It was interesting to note that another teacher suggested a more holistic interdisciplinary approach to teaching about climate change.

> I am not sure how CCE is being defined in MOE. While there is factual information that was given about climate change, they are either too general or decontextualised. You are right to raise the point about getting students to raise questions about issues in climate change because of the complexity of it – it is not simply "global warming" in a homogeneous manner. Different parts of the Earth are experiencing different changes. Another issue which you have raised is about infusing across all

> subjects. I wonder how the ministry could lead in promoting interdisciplinary approach towards CCE. I don't think Geography alone is enough to do justice to CCE.
>
> (Chang & Pascua, 2018)

The comments indicate that while teachers feel that CCE is present in the curriculum at present, it should be made more explicit at upper secondary levels and that it should be taught in other subjects in addition to Geography. While the new curriculum for upper secondary Geography implemented in 2013 has rightly included a full topic on climate change, CCE is still not featured more prominently in other school subjects.

In all, it appears that geography teachers involved in the study were confident in teaching the topic of climate change well. While the self-efficacy of these teachers is high, the question however still remains of whether the teachers have the requisite deep knowledge to teach climate change well.

Case study of teachers' understanding of the topic of climate change

In order to elucidate areas that teacher professional development should tackle in addressing enhancing teacher readiness to teach climate change, a study involving 40 teachers was conducted (refer to Chang, 2014). The focus of the study was not to determine how much the typical teacher would know about climate change, but rather to ascertain "problem areas" that could inform teacher professional development for CCE. The study involved a series of focus group discussions conducted with teachers to discover any inconsistencies in their understanding of climate change and to gain educators' perspectives on how climate change is being discussed in the classroom. The discussions were recorded on an audio recorder and were subsequently transcribed and coded for analysis. The protocol used for the focused-group discussions is included in Table 6.6 for reference.

The participants come from varied backgrounds. While each specialises in one discipline (e.g., Chemistry, Biology, Geography), most teach at least two subjects in any given year. The respondents' teaching experiences range from ten months to 30 years in service.

While the teachers believe Geography is a suitable platform to teach the topic, they also noted that climate change covers only a small part of the discussion of environmental issues such as deforestation, conservation and pollution. The depth and extent of discussion are primarily a teacher's prerogative as the issue only makes a small component in the syllabus document. Indeed, the need for deeper discourse has led one teacher to coin the term "superficial touch" in describing the treatment of climate change topics in the curriculum.

In Chemistry, the teachers identify the topic of Air and Atmosphere as a potential area where discussion on greenhouse gases is sometimes included. They are unanimous in saying, however, that they rarely discuss the topic

Table 6.6 Protocol for teachers' focus group discussion

Perception of climate change

- Are you familiar with the issue of climate change? What is your general "feel" about the issue?
- In your opinion, do you think we (the public) have enough knowledge about it?
- Do you teach about climate change? If so, how?
- In your opinion, how could the discourse about climate change be linked to education?

Knowledge of and attitude towards climate change

1. In your opinion, what is climate change?
2. What causes climate change?
    - Is it really happening? Why do you say so?
        - o Do you have reservations about your stand? (Is it possible that there is also an equally compelling counter-argument to your argument?)
    - Is climate change a natural process? Why do you say so?
    - Is climate change caused by human activities? Why do you say so?
    - What is the greenhouse effect?
        - o How is it different from the enhanced greenhouse effect?
    - How do you think the seriousness of climate change might impact our life? Substantiate your answer
    - How do you think we can slow down the climate change phenomenon?
        - o What are ways to mitigate climate change?
        - o How can we adapt to changes in the climate?

Teaching climate change (curriculum, pedagogy, assessment)

- Where can you get the information/resource you need if you want to teach climate change to your students?
- What sort of PDs would you need for climate change?
- How do you know if your students learn it well?
    - o From your experience, how receptive are students to discussions about climate change?
    - o How are students assessed about their climate change knowledge?
    - o I understand that this is barely an assessable topic. How then do you engage your students to take climate change topics more seriously?
    - o Are they exposed to alternative modes of learning outside of the classroom? (Give the teachers a chance to discuss about school projects/initiatives on environmental education)
- If you were to design a curriculum for your discipline in which CCE is incorporated, what would it look like?
- How, for example, do you envision making the elusive link between knowledge and attitude modification and behaviour change?
- What do you think are the barriers to effective teaching about climate change?
- What do you think are the barriers to effective learning about climate change?

(Source: Chang, 2014 p. 89).

because it is not integrated into the textbook and it is not explicitly included in the syllabus structure as well.

Similarly, the discussion about climate change in Biology is contained within the study of ecosystems and man's contribution to their destruction or preservation. Central to this is the topic of deforestation. The teachers say they do not delve deep into the matter for lack of time and because the syllabus does not call for it.

In Physics, climate change may be discussed in the topic of energy change and the implications of using non-renewable energy sources. However, the teachers acknowledge that their current practice does not include talking about climate change. One teacher even claimed that from a Physics point of view, he does not see many links between the discipline and observed fluctuations in the climate system.

In other disciplines such as History, English, Music and Math, the teachers say they seldom use climate change scenarios in their classes. For instance, English teachers may use news clippings on climate change for reading comprehension. In Math, teachers use PSI readings in their Statistics sessions.

Knowledge about climate change

CAUSES OF CLIMATE CHANGE

Climate change is understood as manifested by changes in erratic weather patterns. These changes happen over a long period of time, around 30–40 years. One teacher mentioned that the climate might also change due to volcanic eruptions. Another explicated that he/she is not familiar with the mechanics and explanations of why it is a natural process and deems it is due to the Earth's magnetic forces and gravity pull between the sun and the planets.

Teachers are aware that climate change is a natural phenomenon that is aggravated by human action. More than natural causes, the teachers are convinced that the enhancement of the GHE is due to human activities. The teachers were able to contrast the natural GHE and the enhanced one. The natural GHE is keeping the Earth warm enough to support life. The enhancement of the GHE mainly through carbon dioxide emissions is man-made. Further, the enhanced phenomenon is seen as an indicator of man's relentless quest to acquire more and more things, of depleting the Earth's natural resources, mainly a reflection of his consumption patterns.

One discussant muses about the current debate in the scientific community on whether the phenomenon is man-made or natural. He also added that evidence shows there's a rather rapid change after the industrial revolution when in fact such changes should take millions of years. Another reiterates this acceleration and highlights the current actions of developing countries in exploiting natural resources and thus releasing carbon into the atmosphere due to the use of dirty fuels. Essentially, the teachers stress that these trends show a state of imbalance on Earth, a deviation from its natural state.

Some teachers expressed uncertainty in determining how weather patterns observed can be attributed to climate change. A number admitted that they may have basic knowledge about how the increase in greenhouse gases such as carbon dioxide aggravated the situation but are not certain of the dynamics and the time frame involved.

EFFECTS OF CLIMATE CHANGE

Teachers associate climate change mostly with changes in temperature. This increase in temperature is due to the trapping of heat. Such is due to the increase in the amount of greenhouse gases, mainly carbon dioxide, in the atmosphere. While the increase in temperature follows a natural cycle in geologic history, the current pattern of temperature increase shows that it is speeding up at a rate that is "a bit too fast than normal".

With heat-trapping comes, temperature rise, referred to commonly as global warming. In general, it will cause the polar ice sheets to melt, sea level to rise and may induce natural disasters such as droughts, floods or more severe storms.

In Singapore, it will mean a much warmer temperature. One teacher gave an example of freak weather that resulted in the flooding in Singapore and Thailand. Thunderstorms will be more frequent as well as sea level rise, which could also lead to coastal erosion. The residents' lifestyles will be greatly altered. As small islands are envisioned to disappear, areas that Singaporeans used to visit may not exist anymore.

The teachers provided examples of such as irregularities in the onset of seasons in other countries, unusual rainfall patterns and wild fluctuations in temperatures recorded. One teacher mused about not buying a first-level unit for it just might be flooded in the future. Another commented about how it will become too expensive to dispose of trash as there might be big issues concerning the use of incinerators in the future. As a nation subsisting on the import of food products, Singapore will feel the brunt of climate change if the crops from its supplier countries become affected. It will then become harder to source food products. Water may also become scarce and expensive.

MANAGING CLIMATE CHANGE

Preparing the young generation to manage the effects of a changing climate by giving them accurate information and equipping them to make a reasonable judgement of their own are some of the most noted management strategies emphasised by the teachers. The most popular, top-of-mind answer on how to mitigate the effects of climate change is the 3Rs – Reduce, Reuse and Recycle. A number of the teachers posit that Singapore's affluent lifestyle deprives its citizens of becoming aware of the repercussions of excessiveness in consumption. One teacher suggested that reducing affluence could help. Opposing this view, another indicated that development or affluence is not the main problem. In the case of Singapore, it is unsustainable development that is the threat. From a wide-angled perspective, a solution such as creating laws to truly enforce the 3Rs was mentioned alongside giving incentives to those who comply with it. The idea is to curb extreme practices which result in excessive waste.

The teachers stress that reversing the phenomenon may not be the way to go. It may be a better idea to slow it down. Overall, concerted effort is the key for any strategy to be successfully implemented. This could only happen if people are aware of their individual efficacies to take on the challenge and take ownership and responsibility for it. In this light, one teacher stated that awareness may not be enough in ensuring that the person on the street acts responsibly based on his awareness of the repercussions of his actions.

In terms of bringing the idea to the classroom, the teachers think that encouraging each student to do his or her part means a world of difference. It is highlighted as well in the discussions that teachers themselves should show the children how to "walk the talk": of turning off the lights when daylight suffices, of regulating the use of the thermostat, of saving on printing, of reducing the use of disposable products, etc.

In general, the interviews uncovered a good range of knowledge, skills, attitudes and actions about CCE in the teachers but there are still some of them who have indicated challenges to teaching CCE effectively. This mirrors findings in studies around the world too (Hestness et al., 2011).

Challenges to effective teaching

Teachers may be unsure about the elements and dynamics of climate change. One admits to being unfamiliar with the scientific mechanism involved, and another claims to not know much about the natural causes of climate change and the extent to which it threatens the sustainability of human life. Some have bits and pieces of information, mostly based on perception, and those gathered from media reports. One Biology teacher even claims that she has merely textbook knowledge that is as limited as those held by her students. They think they might need to verify the truthfulness of their held beliefs. Among the key barriers to effective teaching cited by the teachers are commonly held misconceptions about the topic, concepts being too abstract, the scope being too broad, limited curriculum time, insufficient family and societal support in reinforcing students' learning and that climate change is only one of the many environmental issues that require urgent attention.

MISCONCEPTIONS

Among the most noted misconception held by some of the teachers in this group is that the greenhouse gases form a layer that becomes thicker as humans release more carbon dioxide into the atmosphere, thus, creating the enhanced GHE. In addition, it is also believed that the ozone gets thinner due to pollutants released by oil refineries. It is also believed by some that UV rays cause global warming, and one is of the impression that CFCs are

Table 6.7 Summary of teachers' readiness in CCE

	Causes of climate change	*Impact of climate change*	*Managing climate change*
Knowledge • Recall	Teachers are able to state that the enhancement of the GHE is mainly through carbon dioxide emissions	Teachers associate climate change mostly with changes in temperature	Teachers typically mention the 3Rs in mitigating climate change
• Understand	The teachers are able to contrast the natural GHE and the enhanced one	Teachers are able to explain that the ice sheets in the North and South Poles melt, causing sea level rise and may induce natural disasters such as droughts, floods or more severe storms	The teachers stress that reversing the phenomenon may not be the way to go. It may be a better idea to slow it down
Skills • Analyse	One teacher is able to use the concept of time scale to argue for natural causes of climate change	The teachers provided examples of such as irregularities in the onset of seasons in other countries, unusual rainfall patterns and wild fluctuations in temperatures recorded.	Teachers were able to argue for a more sophisticated approach to mitigating climate change, the scale of analysis used was the individual level, emphasising personal responsibility as a sustainable strategy
• Evaluate	One teacher is able to evaluate the climate change caused by humans using the concept of unprecedented rate of change in temperature	Not uncovered in the interviews	
Values	Teachers are convinced that the enhancement of the GHE is due to human activities	Teachers feel that Singaporeans' lifestyles will be altered by climate change	Preparing the young generation to manage the effects of a changing climate by giving them accurate information and equipping them to make reasonable judgement of their own are some of the most noted management strategies emphasised by the teachers
Action	Not uncovered in the interviews	Not uncovered in the interviews	Some teachers reported that they walk the talk and lead by example in "turning off the lights"

(Source: Chang, 2014 p. 93).

no longer used so that the world is quite safe at its present state. Some are of the impression that heat-trapping is due to the infrared rays that come in and go out. Another states that the sun's energy gets trapped, not the heat from the ground. They also tend to associate climate change with other environmental hazards such as tsunamis, earthquakes and pollution in general.

CONCEPTS ARE TOO ABSTRACT

"It is always difficult to teach an abstract concept", says a teacher, referring to her struggle to enhance students' engagement in lessons. Discussions on this topic were lengthy and teachers took time to mull over the elements that make it hard to connect students' knowledge to real-life applications. A number of teachers expressed their apprehension to teach the topic in depth for fear that they may be preaching the ideas as they themselves doubt the veracity of their knowledge.

The teachers also pointed out that children in Singapore may not see the relevance of what they are learning to their urban lives, when their closest encounter with nature, as one teacher says, is through the television. Some teachers call it a cultural and social mindset, a predisposition to be disconnected from nature because the child most often has no interaction with it. As one teacher illustrated:

> You know what, I brought them to a waterfall. They went there, they stood up, you know. They were just looking, "Teacher, what to do, ah?" I say, "You go down. Go and touch the water". They just don't know how to interact with nature, with the waterfall! And I have to show them, you know. "Look, you do this!".

THE SCOPE FOR TEACHING CLIMATE CHANGE IS TOO BROAD

Climate change is an issue full of conflicting conceptualisations with little room for learning in the formal curriculum. For teachers, this situation does not give them straightforward facts and makes them uncertain of what to teach. One teacher singled out the debate to mean there is a dearth of credible information to base judgement.

In addition, the teachers perceive that there might exist an overlap across disciplines in the discussion of the matter. One teacher commented that secondary-level students may not be expected to know very much about the issue; they are merely expected to have some degree of awareness about it and maybe act on it. She explains that junior college and polytechnic level students are expected to be more receptive and active in these ideas.

INSUFFICIENT CURRICULUM TIME

In itself, climate change is not examinable in some subjects. One teacher said she could teach more about climate change but that would mean it would eat into time meant for other contents. One teacher laments, "At the end of the day, they're sitting for the national exam, and it's results that everyone wants. So it's very difficult. It's not easy". It has come to a point, another teacher added, that there is just too much on the plate of both teachers and students. They are also "pulled in so many ways" by other factors so that time is an important element that both teachers and students do not have. As one teacher stated, encouraging the students to think broadly and deeply through providing additional materials for analysis is a challenge. What then happens is simply "a test [of] just memory". In some cases, a teacher may skip detailed instruction when pressed for time.

One teacher aptly illustrates the point that lack of time prevents her from fully engaging students in widening the breadth of investigation of conflicting perspectives and as well as in deepening students' analysis of trends. She would gladly learn with them, for example, the projection rates of temperature increase and the elements scientists put into the equation for these projections.

TEACHING RESOURCES

Teachers base their instructions on textbooks that developed with the curriculum cycle and are updated around every five years. They supplement data from various sources, among which the most cited are the Internet, newspapers, documentaries such as The Inconvenient Truth and clips from the site YouTube. Others refer to research papers and content knowledge they obtained from their university studies.

The intensity and sustainability of teaching relating to the environment in general, and climate change in particular, are also bolstered by school leaders' support and teachers' collaborative efforts. In one school, the teachers state that the principal's pro-environmental program such as the "Green Day" helped them sustain teaching through interactive activities with the students. They also practice collaborative, collegial discussions with colleagues who may know a lot about the subject. Some also engage in Professional Development (PD) activities such as seminars and staff meetings. As one teacher said,

> The sharing between the teachers, where we sat down and plan the lessons, and hear how the different Sciences actually approach the matter, was very interesting. Yeah, I mean, you cannot do it as an individual. There needs to be collaborative efforts across the subjects.

One product of such collaborative activity is an initiative called "Climate Change Crusaders" in one of the schools. As a start, the teachers in this PD program selected students to be part of this activity to further inquiry in the area, guiding them accordingly in the areas of Biology, Chemistry, Physics, etc.

RECEPTIVITY OF STUDENTS TO THE TOPIC

The teachers say students are mostly aware of climate change due to the proliferation of information on the matter. However, students' understanding may be sketchy, and they may be confused with some terms and processes involved. One teacher, for example, noted that students in her class view the GHE to be bad for the Earth.

Compared to other topics (e.g., Racial Harmony), one teacher observes that climate change interests students more. In one instance, a group of teachers also shared their experience of a facilitated discussion with a group of students about climate change that lasted for two hours. The interest, according to teachers, is present. It just needs to be nurtured.

On the other hand, teachers also hear students complaining about the tediousness of doing projects as well as their disinterest to study beyond passing the exams. One teacher shared an instance wherein the students asked her to skip the topic altogether because they claim to have learnt about it already in another subject. In addition, teachers opined that they may not be sure whether all things taught to students are taken seriously, considering that central to the discussion on climate change is focused on temperature changes around the world and not on their local context. Still, another commented that it may not be a lost cause at the end of the day because as teenagers become more mature, they tend to become more receptive to these issues and what they learnt from school provides a good foundation.

LACK OF FAMILY AND SOCIETAL SUPPORT IN REINFORCING STUDENTS' LEARNING

The teachers note that whatever is learnt at school is not reinforced when the student steps out of the classroom, goes home and deals with the norms of society. In essence, teachers point out the inconsistency of practices by the general public with what is deemed ideal behaviour. Some conveyed their experience of encouraging kids to practice the 3Rs, for example, only to be deterred by the unwillingness of family members to follow suit.

The teachers are aware that the society they are living in reflects extravagance and wastage and this is manifested in all facets of their everyday lives such as in the way people consume and accumulate goods, in nationwide celebrations and events such as the F1 race, the ubiquity of air-conditioned spaces and well-lit highways. The act of going against these norms may prove to be an inconvenience to a young person.

CLIMATE CHANGE IS JUST ONE OF MANY CONCERNS

One teacher explicated that climate change is but one of many issues confronting the world today, and essentially, one topic that must gain relevance and inclusion in school learning. For example, she said that she is worried more about the lack of coverage of plate tectonic movements in Geography classes which she thinks is more immediate as she views climate change to take 10–20 years.

In the classroom, climate change does not figure as a priority topic. A Biology teacher refers to the discussion on climate change to be "touch and go", meaning, they study deforestation and the carbon cycle which may or may not include touching on climate change. The complexity of the issue, as to the causes and effects and the many uncertainties posed by scholars, deters teachers from expanding instruction about the matter. They are aware that the discourse is evolving and they are wary that they might be teaching something that will be totally changed the next day.

EFFICACY OF ASSESSMENT

The seriousness upon which any topic is considered is based on whether it is assessable, in particular, if it is part of a high stakes examination. In the case of climate change, it is rarely so. As to the efficacy of assessment, teachers indicate that they are of no value if learning does not translate into action. Linking the two, according to educators, is the most difficult task for them. Test scores are not reliable indicators of whether a person cares for and acts for the common good of other persons and of the Earth. As one teacher mused, "Action is different from tests and scores, and action is very difficult to measure actually". Then again, one teacher also laments that even as she tries hard to connect concepts with action, and students may actively participate in it, at the end of the day, the learning may not be that much and is simply "thrown out of the window".

Suggestions on how best to teach climate change

BE WARY OF DUPLICATION

There may be repetitions in the teaching of concepts to students across the various disciplines. For example, students may learn about the causes of climate change in varying degrees in Chemistry and Geography. This duplication may result in students switching off. Thus, there should be synergy among the subjects. More importantly, the disciplines should see climate change from a holistic perspective and not teach it in a segmented way.

WALK THE TALK

Teachers acknowledge that they are at the forefront of learning, the main drivers of change due to their authority of being knowledge imparters and thus, the need to live responsibly as they preach to students. Teachers must make use of critical moments of showing their students that it is possible to practice the 3Rs, for instance. One teacher shared her habit of using her own reusable container, for example, to buy food from the school canteen. This way, she is able to show students that it is possible to do away with disposable products. The same goes for using both sides of a paper, being conscious of using less energy when possible, through switching off the air conditioner or an extra light in the faculty room.

DO NOT TEACH JUST FOR EXAMS

While they are aware that the stake depends on the weightage of the topic in the exams, the teachers emphasise that a person's understanding of climate change is relevant beyond the walls of the classroom and is necessary for the individual to function as a responsible citizen of the world. They are aware as well that students themselves may not necessarily study for the sake of understanding but merely to score well in the exams, which is a habit that does not translate to meaningful results such as attitude change. As such, they stressed that they must not be constrained by this same structure – to not rush lessons to complete the syllabus outline, to simply regurgitate the content of textbooks. In line with this, they believe that climate change should be taught in such a way that it is relevant to students' present realities. Field studies, museum visits, immersions and experiments are some of the possible suggestions to bring the concepts of climate change closer and more tangible to the consciousness of students. Some teachers complained that it is not easy for them to take students to local sites such as Pulau Semakau or the Gardens by the Bay due to the fees charged.

LOOK AT THE FACTS CLOSELY

Some teachers indicated that they are not very confident of the extent of their understanding of climate change concepts. They are then aware that they must boost their content knowledge. One indicated that teachers must take notice of the facts well. In an example, he explained that global warming has become the poster boy of climate change when in fact the phenomenon may have an opposite impact as well, such as global cooling. The teacher suggested that teachers should proceed with caution and that educators must be provided with "a better resource on how to break down big ideas into something very bite size for children to understand".

TRUST YOUR STUDENTS

It is put forward by the teachers that exposing students to experiments and projects will help them infer and develop their own theories on the elements that cause climate change, its effects and how to manage it. Through hands-on activities, learners are able to collect, analyse and interpret their own data, do away with an erroneous understanding and validate reported claim –thus, authentic learning. Through these micro-experiments, experiential learning and simulations of real-life events, students are allowed to form a firm grounding in their knowledge base and empower them as learners. It is also emphasised that children should be allowed to reflect and discover themselves, not forced into it.

BE RELEVANT

One of the key features of climate change learning that is identified to deter meaningful understanding is its perceived irrelevance to the local context. Footages gathered through the Internet, for example, feature extreme scenarios that seem to resemble little or none of the Singapore situation. Teachers then expressed that they should fine-tune their lesson plans in consideration of the local setting. It is then put forward that a lesson package would be helpful with a set of video clips, experiments and worksheets contextual to Singapore.

In addition, one teacher explained in length the need to be in tune with what students' opinions are, their current understanding of the phenomenon, the values they hold towards the idea of climate change and their level of interest. From these elements, a teacher can then infer the appropriate strategy to bring in the topic of climate change, rather than going through the contents of the syllabus haphazardly.

ALTERNATIVE MEANS OF TEACHING

Apart from lectures and classroom discussions, the teachers also shared alternative modes of learning about climate change such as through competitions, paper proposals and research. One teacher brought her students to investigate an industrial partner's range of eco-ideas initiatives. She was also exploring opportunities with other industrial partners to collaborate with to fund the students' projects. Others use events such as the Earth Hour to stage awareness campaigns with students and bring in experts to coach the students on the how-tos of living urban and environmentally responsible lives, for example. This is an avenue as well to scale up the curriculum and add substance to the content of the syllabus.

Environmental (green) committees, clubs and working groups are sprouting in some schools, which are meant to scaffold student learning with real-life applications. It is envisioned that learning becomes sustained as students are exposed to activities of these groups more frequently than lectures.

CURRICULUM ON CLIMATE CHANGE

As the study was conducted just before the implementation of the 2013 upper secondary Geography syllabus which features climate change within a key topic of study, teachers feel that the syllabi in the various disciplines that touch on climate change do not give much breadth and emphasis on the topic. As teachers depend much on the structure of the curriculum and the content of the syllabus, there is not much that they could do in terms of adding more discussion time or more coverage in terms of assessment.

When asked as to the changes they envision in the teaching of climate change if they have complete control of the curriculum, the teachers indicated several points of interest. On whether the curriculum serves the purpose of prompting students to be responsible inhabitants of the Earth, one teacher countered that understanding is not the main problem as students may churn out information and argue impeccably but may not necessarily be responsible enough to alter their lifestyles or habits. The syllabus and the curriculum only deal with what should be covered in the exam. Action, in this sense, is left out. It is put forward that students are already bombarded with more than enough information from the media. The next obvious step would be to link what they already know to what they could possibly do to help the situation. Beyond awareness and knowledge building, the teachers are concerned about impacting students' lifestyles.

Several favour a curriculum that is action-oriented and complemented with content that is relevant and engaging for the learner. In this sense, students are encouraged to dig deep through guided instruction. They will be given the rein to investigate the complexities of the debate and the nuances of arguments put forward by stakeholders and come up with a mode to determine the extent of understanding of the students. For instance, a teacher said she would encourage students to engage in a long-running project, the aim of which is to allow students' knowledge about the issue to become ingrained and linked with how they view the problem could be solved.

The idea that came is for students to do service learning. As they go through the process, it is hoped that the students could get some value out of it, which will then make a mark on their habits and those of the people around them. A problem-based learning is put forward as a model for service learning by another teacher, a mode in which the student is to find and flesh out the roots of the problem himself for him to be fully convinced of the issue's gravity. Only then he will innovate and come up with his own solution to it.

The idea of partnering with organisations is suggested in the form of sending students as interns to agencies such as the National Environment Agency, National Parks and the Public Utilities Board or even the local media outfit MediaCorp™. They will then be encouraged to ask questions while being exposed to the works of these agencies. Another teacher stressed an interdisciplinary collaboration to bring the matter closer to the consciousness of the students. She believes that students do not have a deficit of information but are generally lacking concrete conceptualisation on how to connect knowledge to practice. Thus, an exposure to areas such as factories, trips and discussions in areas that used to be forested would greatly enhance the process.

Similarly, collaborations in the education setting are also favoured by the informants to avoid the duplication of resources and coverage. If climate change is strategically taught according to the tenets of the Natural and Social Sciences, then education becomes more targeted and the teachers could free up some time in the curriculum for the students to execute what they have learnt.

Some teachers mentioned that they need to upgrade their content knowledge about climate change for them to effectively and confidently teach about it in the form of a professional learning activity or through staff training. Some teachers suggested a "climatic change package", a teaching kit on the essential content and strategies they could employ such as video clips, experiments and worksheets. On the other hand, one teacher commented that content is not the issue as everyone, teachers and students, are bombarded with information already. Still, another commented that it is possible to broaden and deepen the treatment of climate change provided this will not add up to teachers' already very heavy workload.

All the teachers that were participants in the focus group discussions held beliefs and provided suggestions that are truly heartening. In addition to identifying concerns that threaten the effective implementation of a CCE curriculum, they offer suggestions to improve instruction, and weigh in on the efficacy of assessment; the teachers in this group reminded us the need to remain relevant, and challenge behavioural change in the student, so as to achieve a more sustainable future.

A common suggestion among studies around the world and from this focus group of 40 teachers is to provide professional development opportunities for teachers to clarify misconceptions and share best practices.

Teacher education for CCE

From the studies reviewed and through the focus group discussion of a group of 40 teachers from Singapore, the key challenges to teaching CCE well are primarily focused on mastering the knowledge to teach about climate change, developing relevant and efficacious instruction, engaging student interest, adopting effective assessment and having sufficient and effective support for professional development.

The key avenues through which the issues of deep understanding of knowledge and pedagogical practice can be addressed seem to be through teacher education through professional development. There is also a need to address teacher education at the pre-service level. Hestness et al. (2011) examined a curricular module on climate change in an elementary science method course. The teacher trainees were taught the following: how to engage students using authentic scientific data, how to incorporate technology into classroom discussions while adopting global perspectives on science and the different forms of assessment on student understanding of climate change. In addition, teacher trainees also researched and discussed topics related to climate change. The results showed that there are positive impacts on the teacher trainees' knowledge on "climate change, confidence to teach, and awareness of resources to support their future teaching" (Hestness et al., 2011). However, there are no follow-up actions on the teacher trainees, whether whatever they have learnt in the teacher preparation course have been implemented in real classroom teaching. As such, evaluation of the effectiveness of this inclusion in teaching climate change in the classroom cannot be carried out as well.

Thus there is a need to carry out studies to find out what and how climate change should be taught in pre-service and in-service teacher education to prepare teachers well to teach climate change to *help learners develop knowledge, skills, values and action to engage and learn about the causes, impact and management of climate change*. In addition, there should be follow-up studies on whether teachers were able to implement what they learnt onto real classroom teaching. Furthermore, the effectiveness of such programmes to improve their teaching should be evaluated on the basis that they have implemented what they learnt onto real classroom teaching.

Teacher preparation for CCE

Several approaches have been suggested in the literature on preparing teachers for the teaching of climate change. In one example, there was a focus to equip teachers to disseminate accurate information and instructional strategies to help students strengthen their skills in critical thinking on the topic of climate change (Mckeown & Hopkins, 2010).

While teacher competence remains a key issue across international studies, teachers must be considered as opportunities to improve the teaching of climate change and not as major contributors to the issue (Schreiner, Henriksen, & Hansen, 2005). To begin with, it is important to understand teachers' and students' prior knowledge and attitudes before embarking on any programmes to improve teacher readiness for CCE. Indeed, CCE should help students understand the complexity of climate science, enable him or her to have an open mind and adopt a critical approach to constraints and management strategies so that they can debate and decide on socio-scientific issues presented to them (Schreiner, Henriksen, & Hansen, 2005).

Further, CCE should endeavour to encourage responsible attitudes while avoiding propaganda and moralisation. Currently, CCE suffers from a limited scope and few are inclusive enough in terms of the scientific, societal, political, ethical and personal discourses. CCE should empower the learner to appreciate that he or she has a stake in how the future will turn out (Schreiner, Henriksen, & Hansen, 2005).

Teachers commonly focus on changing human behaviour to mitigate the impact of climate change when they think about how to teach it. From the review of studies carried out around the world and the responses from teachers in Singapore, few are able to articulate the best ways to teach a topic on climate change, considering the subject matter and pedagogical knowledge. Indeed, a teacher must first have a grasp of *what* and *how* a topic should be taught in order to develop a robust approach to teaching about climate change.

Determining what to teach for CCE

As part of a research study and in fulfilling the role of a teacher educator, I developed an in-service workshop for geography teachers. The workshop was designed on two key principles. First, teachers uncover their own subject matter understanding through concept mapping and secondly, to explore how climate change can be taught by examining the conceptual lens through which CCE can be framed.

An expectations-setting activity was conducted through email correspondence before the commencement of the workshop. This activity uncovered several areas of concern, including learning how to be able to introduce climate change as a topic, how to engage students and helping students to recognise the large-scale impact of climate change.

The workshop was designed to engage curriculum subject matter for climate change, grounded in cognitive engagement theories. In particular, the proposed taxonomy of learning outcomes in the dimensions of domains of knowledge, cognitive engagement and the aspects of climate change provides a vocabulary that can describe what a student has learnt about climate change. Consequently, we have a framework with which to describe and determine how CCE can develop a more informed community that can better manage climate change issues.

In other words, we can subsequently use this framework to describe the learning outcomes of any climate change learning task. For example, whether learning has simply resulted in students recalling turning off the fan before leaving a room or being able to evaluate and critique a newspaper report on climate-change-related flash floods. Concepts of climate change, global warming and the GHE are commonly misunderstood (Lee et al., 2007; Cordero et al., 2008; Hansen, 2010; Francis et al., 1993; Koulaidis & Christidou, 1999). Indeed, there is often no distinction between learning about the science, impact or management of climate change. To this end, the subject matter knowledge of climate change can be organised around the macro concepts of causality-consequence-management. Firstly, it is imperative that students understand both the natural and anthropogenic causes of climate change, before learning about the impact climate change has on human lives.

Consequently, mitigation or adaptation strategies can be learnt in the context of the types of climate change impact. To this end, the workshop was structured with a few key activities to achieve the aims above. Firstly, the teachers in the workshop had to advance a definition of climate change. Questions for clarifying teachers' gaps in knowledge or misconceptions were invited throughout this brainstorming phase. After the clarification phase, participants had to work in groups to develop a concept map of the various concepts brought up during the brainstorming session. The maps were then examined and each group had to represent their understanding of climate change as a definition.

Activities of the workshop

As teachers have pointed out in the focus group discussions, the scope of learning about climate change is wide and there exist several misconceptions on the topic. Teachers were tasked to examine the topic of weather and climate, focusing on the definitions and the key concepts provided in the syllabus document. The participants then took part in a brainstorming activity to determine what it means to teach climate change based on the conceptual framework of CCE. Several guiding questions were asked to help teachers along, including:

1. What is the difference between weather and climate?
2. What do we mean when we say that the climate has changed?

3. What are the causes of climate change? Are they due to human activity?
4. How does climate change affect humans?
5. What can we do in response to this impact?

Teachers were encouraged to ask questions to clarify or affirm what they understand about climate change throughout the brainstorming session. For example, some of them asked,

1. I read about the conspiracy that green-technology industry has fabricated the claim that climate change is natural. Is this true?
2. Is it true that CO_2 is the only Greenhouse gas that matters? I read somewhere that the other greenhouse gases are insignificant in their contribution to global warming
3. How do I explain to my students that Ozone is not a greenhouse gas?

Consequently, the related subject matter to the questions was explained and clarified. The answers to the three questions are included here to illustrate how they will help clarify teachers' understanding.

1. Regardless of whether the green-technology industry's stand has been motivated by hidden agenda or not, there has been episodic climate change throughout Earth's geologic history. This fact was established long before green technologies came into existence. During the Little Ice Age of 1350–1850 AD, the climate was much cooler than today, for instance. Hence, climate change is a regular feature in Earth's geologic history. A relevant question rather will be whether human activity has diminished or aggravated the rate of climate change.
2. While CO_2 is a main contributing gas to GHE, its contribution is not as significant as water vapour. Further, the other natural and man-made greenhouse gases add on to the GHE. With the addition of all these gases and the introduction of new greenhouse gases to the Earth's atmosphere, we are dealing with an enhanced GHE. So, the question really is what has caused this increase in both concentrations and types of greenhouse gases.
3. Ozone is a greenhouse gas. The confusion students have is between the Ozone that is found in the stratosphere (about 25 km from the ground) and the ozone that is found near the surface of Earth (in the troposphere). Stratospheric ozone forms a protective "layer" that blocks out harmful UV radiation. Industrial activity and photo-chemical activity cause Ozone to form in the troposphere, which then acts as a greenhouse gas. It is important to distinguish these two as the confusion stems from the topic of Ozone depletion in the stratosphere.

The teachers then went on to draw a concept map to organise the concepts that have clarified through the brainstorming activity and the question-and-answer session. The following shows an example of a concept map generated.

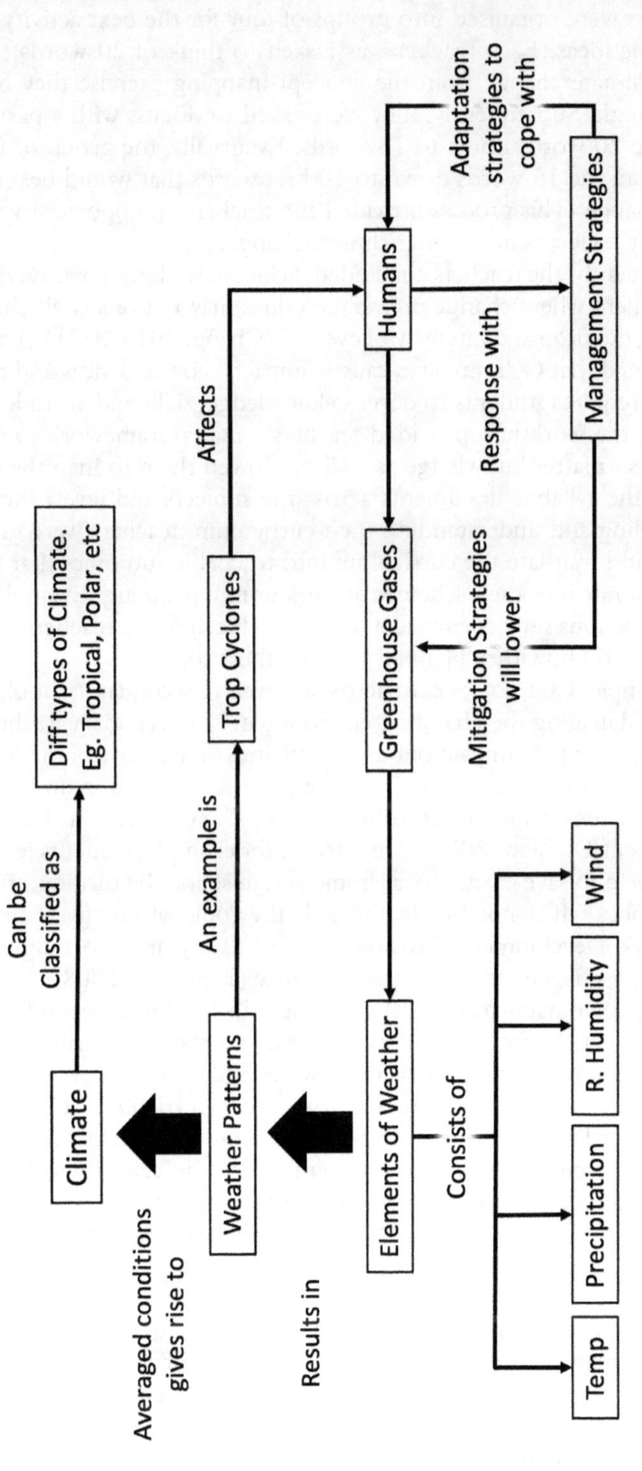

Figure 6.1 An example of a simple concept map of the topic.

Teachers were organised into groups of four for the next activity to consolidate the ideas. Each teacher was tasked to think of 20 words that best describe climate change from the concept mapping exercise they have just gone through. Subsequently, they were asked to discuss with a partner and reduce the 20 words down to 15 words. Eventually, the group of four will further distil the 15 words down to 10 best words that would best describe climate change. This process provided the teachers an opportunity to identify the key concepts in defining climate change.

Consequently, the teachers concluded, "climate is a long-term average state of atmosphere where change in average values may not reflect all changes in frequency, duration and intensity of events" (Chang, 2013, p. 31). The teachers also noted that CCE involves causes, impacts/consequences and management and requires students to develop knowledge, skills and attitudes.

Indeed, the workshop provided teachers a macro-framework to organise their subject matter knowledge as well as allowed them to map the relevant topics to the syllabus documents across the subjects and levels they teach. From reading and understanding their curriculum, teachers were guided to interpret and translate the curriculum into teachable subtopics that teachers can incorporate into their schemes of work or unit planning. Exemplars from across the various subject curricula were introduced to the teachers who then commented on how they planned to teach these topics.

For example, CCE topics exist across all levels in secondary school. On the topic of "Managing the changing environment" in Secondary 2, the curriculum focuses on the "impact of human activities on the environment at local, regional and global scales, justifying the need for protection and conservation of the environment at different levels" (Curriculum Planning & Development Division, 2005a). In science, for example, students learnt how they can help to save electricity at home and in school by turning off electrical appliances when not in use through the topic of energy (Curriculum Planning & Development Division, 2007b). They are also taught to use more energy-efficient appliances, for instance (Heywoth, 2008).

"Caring for our environment", is a topic in the lower secondary Social Studies subject. It requires students to explain the causes and impacts of environmental problems such as land, water and air pollution and describe effective ways to manage the environment (Curriculum Planning and Development Division, 2006). In addition, there is the inclusion of climate change in the curriculum where students learn the causes and impacts of global warming as well as to describe and evaluate the management measures adopted. Finally, the topic "Sustaining Singapore's economic development" in the upper secondary Social Studies subject focuses on the concept of balancing economic development with environmental management.

However, there is a need to help students connect the concepts of causes, impact and management of climate change. Some of the questions used to guide teachers in helping students link the concept are as follows:

1. When we talk about climate change in the last 30 years, what are the changes to the weather patterns? Are the changes the same across regions?

2. What has caused the climate to change?
3. How do the changes in climate affect the amount of water in the hydrologic cycle? Does it affect the amount of water in the oceans?
4. How will changes in climate affect plant and animal life? Do these changes vary across regions?
5. How do changes in climate affect our everyday lives?

Teachers should first understand the impact of climate change on human lives and the environment before getting students to evaluate different climate change adaptation and mitigation strategies. While adaptation describes strategies to live with climate change, mitigation refers to doing something to ameliorate the causes and impact of climate change. While both management strategies enable people to respond to the climate change issue, the nature of management must be elucidated to help students analyse the strategies and evaluate if they are effective. Indeed, some of the suggestions that came out of this discussion were to use graphic organisers to help students "structure their evaluation" (Chang, 2013, p. 32).

In addition, the participants were encouraged to extend students' cognitive engagement of specific knowledge domains. For instance, in addition to considering the domain of impact and mitigation, the teacher should also help the learner build skills to evaluate and take action that would be congruent with the recommendations of climate change mitigation. In essence, a teacher should organise and plan a lesson with the scope and mode of assessment in mind so that the students' learning can be used to indicate if the individual has learnt the requisite skill and knowledge.

Set against a social landscape that is concerned about the need to respond to climate change (Hussain, 2007), CCE has become increasingly important in many countries. The first workshop conducted helped teachers articulate their subject matter knowledge through a conceptualisation exercise. The deliberate focus on subject matter knowledge was informed by similar recommendations in studies around the world and from Singapore. However, this workshop is still largely focused on the "what to teach" rather than the "how to teach". A second workshop conducted about a year from the first was focused more on the latter and uses the Conceptual Change framework which was then used in the Refutation Instruction introduced in Chapter 5.

Conceptual change as a means to teach about climate change

In considering the work of Duit & Hewson (in Treagust & Duit, 2008), a professional development workshop to help teachers uncover "how to teach" climate change should start by correcting teachers' content knowledge first, then progressing on to collaboratively restructuring students' pre-instructional understanding, making use of the conceptual change approach. Conceptual change refers to the process in which the learner "moves from not understanding how something works to understanding it" (Mayer, 2002, p. 101). By systematically facilitating cognitive conflict through refuting the errors in a learner's set of pre-instructional beliefs, it is deemed that accurate knowledge will then gain more weight and acceptability to replace old, personal theories. Stepans (2003, p.8) puts forward the Conceptual Change Model (CCM), composed of six stages:

1. **Commit to an outcome**. Learners are made aware of their preconceptions before an activity starts
2. **Espouse beliefs**. They are then asked to share their personal theories with their peers
3. **Confront beliefs**. Through further discussions, experiments and tests in small groups, learners are encouraged to deal with inconsistencies in their beliefs
4. **Accommodate the concept**. Learners are directed to resolve the conceptual conflicts encountered in Stage 3. They are encouraged to assess their prior knowledge vis-à-vis the results of an experiment or discussion
5. **Extend the concept**. They are then asked to identify connections between what they learnt in class and their daily lives
6. **Go beyond**. Learners are encouraged to explore situations related to the concept and extend inquiry by asking more questions

For the teacher professional development workshop, the teachers were taken through the exact same six steps of the CCM. The intent was to familiarise the teachers with the framework of CCM and achieve the goal of "learning by doing" (Lesgold, 2001).

Conceptual change for teacher professional development

With due focus on correcting erroneous knowledge, Boyes, Chuckran & Stanisstreet (1993) encouraged educators to work at disentangling various issues and sources of confusion but cautioned against the tendency of espousing a non-holistic approach. The incorporation of students' alternative conceptions and pockets of learning difficulties into curriculum materials and the professional development of teachers is similarly espoused by (Lee et al., 2007). Gowda et al. (1997) suggest that the collaboration between scientists and educators is paramount in correcting misconceptions through "innovative curricula" (p. 2239) of fine-tuned concepts, up-to-date design and materials. Indeed, "getting to grips with the climate problem will require present and future citizens to make ethically founded knowledge-based decisions" (Schreiner, Henriksen, & Hansen, 2005, p. 41).

The following section outlines the teacher professional development workshop that was conducted with conceptual change as a framework. The discourse will be presented as a narrative so as to provide a holistic description of the proceedings.

The idea of conceptual change was introduced as a "[r]epair process in correcting students' naïve knowledge about a domain" (Chi & Roscoe, 2002 p.3). However, the process is difficult as students lack the awareness of when a change in their understanding is necessary. In most cases, they lack an alternative category to shift into. In other words, they must be able to shift their naïve knowledge into a formally accepted body of knowledge. If students hang on to their naïve knowledge, it will present impediments to deep understanding. Indeed, misconceptions can be highly resistant to change and persists strongly even in the face of the most ingenuous forms of instruction (Chi & Roscoe, 2002). Perhaps the most frustrating aspect of misconception is that students often do not know about their own misconceptions. Plainly, students do not know what they don't know.

Misconception can occur are various levels. At the proposition level, the declarative knowledge can be flawed. These are often manifested as sweeping statements that are wrong such as "greenhouse effect is caused by the hole in the sky". Misconceptions can also occur at the mental model level, which is further described by whether the understanding is coherent, accurate or complete. Misconceptions can be incoherent such as the inability to link cause and effect. Mental model misconceptions can also be inaccurate such as thinking that the heat trapped by the GHE is reflected sunlight from the ground. Mental models can also be incomplete in that parts of the argument may be missing in trying to explain how the GHE works.

The teachers were then presented with examples of how the causes of climate change could suffer from misconceptions at the propositional level. A list of misconceptions found from the literature and my own study was presented:

1. Ozone depletion causes the air to heat up
2. CFCs are the main cause of global warming
3. Climate change is mainly human-induced and not natural
4. Nuclear activities are responsible for global warming
5. Air pollution causes global warming
6. Chemicals used in industry cause global warming
7. Businesses (not industry or transportation) are responsible for global warming

Each of the misconceptions listed was discussed at length, with the teachers seeking clarification from the trainer. The next set of discussions focused on misconceptions about the effects of climate change at the propositional level:

1. Climate change causes Tsunami
2. Climate change causes Acid Rain
3. Climate change causes Earthquake
4. Climate change causes Skin Cancer
5. Singapore is safe from climate change as it is shielded by other countries
6. Climate change has an effect on Singapore
7. Climate change is easily perceptible – for example, "It is hotter than one hour ago"
8. Singapore will have four seasons

This is followed by a discussion focused on a set of misconceptions about the management of climate change at the propositional level.

Management of climate change

1. Reduce littering to reduce climate change
2. Limit the use of CFCs
3. Do not use nuclear power
4. Turn off the lights (no understanding of reducing emissions through reducing power production)

Misconceptions can also occur at the Mental Model level. The following example was used to illustrate how mental models may be flawed because they are incoherent:

STUDENT: Global warming will cause Tsunami
TEACHER: Can you explain to me, how?
STUDENT: Hotter temperatures cause ice at poles to melt. Melted ice means more water in the oceans (pause)
TEACHER: And then?
STUDENT: Too much water in the ocean will be pushed to shore during earthquake. Actually, I am not really sure.

Some mental models may be coherent, and their assertions are correct, but the organising concepts may be flawed:

STUDENT: Heat is trapped within the earth
TEACHER: Where is this "heat" from?
STUDENT: The sun. Heat from the sun's rays. Because of the hole.
TEACHER: What hole?
STUDENT: The ozone hole. Because of too much CFCs we have the ozone hole. So because of that, more sun rays enter the earth which causes more heat. This is trapped by the greenhouse so the heat cannot escape into space.

Finally, mental models may not be competed, in that it has the majority of key propositions but are not completely correct or coherent:

> We must reduce our carbon footprint. E.g. We must refrain from using air con. We should use fan instead. Air con use CFCs which is a greenhouse gas. Greenhouse gases trap heat thus exacerbating global warming.

In the last example on an incomplete model, most of the arguments are right except that the fact about CFCs was not well understood. Sometimes, the model may be very basic but correct but is still not complete, as it does not include key concepts to explain the assertions. For instance,

> We must conserve electricity to reduce our carbon footprint

At the domain theory level, confusion among students arises when the propositions are incorrect in the domain, they are discussing but correct for another domain. This is commonly true in students' confusion between the GHE and the ozone hole. In a sense, the concepts are flawed at a domain level. To correct this, it involves a conscious shift of the students' understanding at the domain level. Table 6.8 shows how the concepts can be shifted across domain levels to help students clarify their misconceptions.

In this method, the propositions are first written and students are guided to shift the flawed concepts, which may be incoherent, incorrect or incomplete into the correct domain. After clarifying commonly held misconceptions found in the literature and from previous studies in Singapore, the teachers were taken through the document on the topic of "Variable Weather and Changing Climate" in the Secondary 3 Geography national syllabus.

After this, the teachers were asked to critique and construct lesson ideas that will enable them to teach the topic meaningfully and change changing students' misconceptions. Three examples of lesson plans, each on the topic of the causes, impact and management of climate change were used. In particular, they had to identify any possible misconceptions and critique the way the lesson was planned. They were also asked to develop alternative ways to teach the subject matter.

To illustrate how this was done, the topic on the causes of climate change and how the teachers critiqued it is presented here. The initial lesson idea presented to them is as follows.

This lesson aims to improve students' understanding of the anthropogenic factors that enhance the GHE such as the burning of fossil fuels, agriculture and cattle farming and deforestation. The lesson involves dividing the class into groups using Google Earth as a tool. The general outline of the lesson includes introducing students to the natural and anthropogenic causes of climate change and the types of greenhouse gases and their effects on the climate system.

To begin the lesson, the students are asked to reflect on the question "How do human activities lead to increase in greenhouse gases and cause climate change?" The teacher should lead the students to discuss the activities of deforestation, burning of fossil fuels and land use changes. The lesson then focuses on deforestation and how it contributes to the increase in greenhouse gases.

A video showing the extent of deforestation in recent years will be followed by a Google Earth activity to identify the locations and the areal extent of the problem. Next, the students will be guided to think about the connection between deforestation and CO_2 accumulation. Through analysing graphs and maps, students will be asked to report on their analysis of how the reduction in forest cover can have a direct effect on the amount of carbon dioxide in the atmosphere. Finally, the teacher asks some students to share a key learning point about deforestation as an anthropogenic factor of the enhanced GHE. The students will also be asked to write a reflection log of a key learning point in the class discussion.

Table 6.8 Domain level misconceptions and the clarified concepts

	The greenhouse	*The ozone layer*
Misconceptions	Because of too much CFCs, we have the *ozone hole*. So because of that, more sun rays enter the Earth which causes more heat. This is trapped by the greenhouse so the heat cannot escape into space	Because human activity results in the **increase in greenhouse gases** like CFCs, causing more harmful UV rays to reach the Earth's surface
Concepts shifted	Because human activity results in the **increase in greenhouse gases** like CO_2, there is more infrared leaving the Earth being trapped so the heat cannot escape into space	Because human activity results in the increase in *ozone-depleting gases like CFCs*, the ozone hole is formed, causing more harmful UV rays to reach the Earth's surface

The teachers' reactions to this activity were varied but most of the 15 teachers present offered alternatives to the lesson activity. They felt that the activity offered little in terms of explaining how greenhouse gases contribute to the GHE. Instead, the lesson in its current form only enables students to correlate the concentrations to the observed phenomenon. Bearing in mind the confusion students have about ozone depletion and greenhouse warming, the teachers suggested that the activity be modified so that students could actively shift their conceptual understanding by explaining the difference between natural and enhanced GHEs. This could be done by drawing an annotated diagram to show the difference between the natural and enhanced GHE. The teachers also suggested conducting a similar exercise to that in Table 6.8.

In response to the teachers' critique and suggestions, I then taunted the teachers to produce the expected annotated diagram they want the students to draw. The teachers worked in their groups to produce the diagrams and Figure 6.2 shows an example of the artefact. At first glance, all the diagrams produced by the groups did not look wrong but there were several possible misconceptions about the drawing that were pointed out to the teachers. Firstly, there were teachers that drew the GHE within the analogy of a fish tank. A box is shown in which the various incoming and outgoing rays are drawn for both a natural and enhanced GHE, often the difference between the two are the placement of "fish tank ornaments" like factories and cows to show the enhancement. There is no distinction made between the radiation balance and the increased concentrations of greenhouse gases. Figure 6.3 shows this example.

Secondly, one group produced a diagram of an actual greenhouse – a glasshouse. This presents a conceptual issue as students will think of the effect as being produced by an enclosed space created by solid material. The GHE is

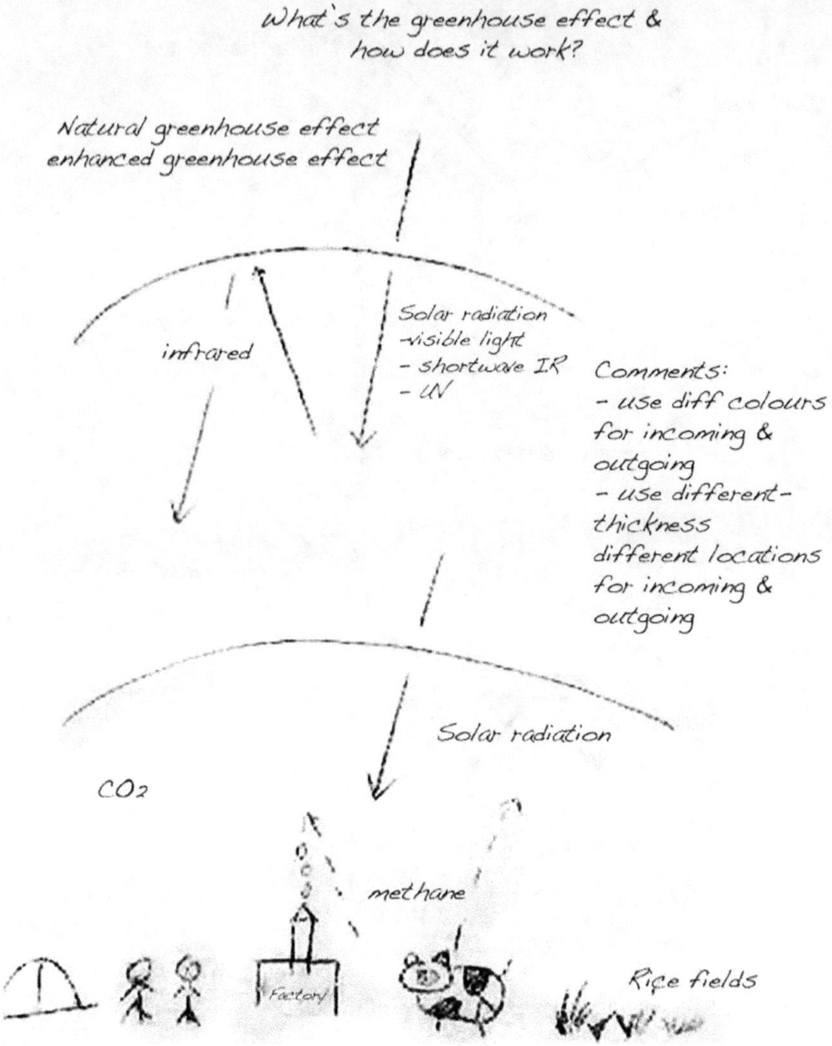

Figure 6.2 Example of the glasshouse analogy used to explain the GHE.

a planetary phenomenon and there is heat transfer in terms of advection and convection. This representation is problematic.

Another typical representation of greenhouse gases is a blanket of gases that envelopes a curved surface of the Earth (only a segment). This representation is also commonly similar to one, which represents the gases like layers of an onion. The problem with this representation is that students will be misled to learn that greenhouse gases only occur as layers at discrete distances from the ground. This is not true. Greenhouse gases are gases, and gases, while they tend to be concentrated in certain altitudes, do not have

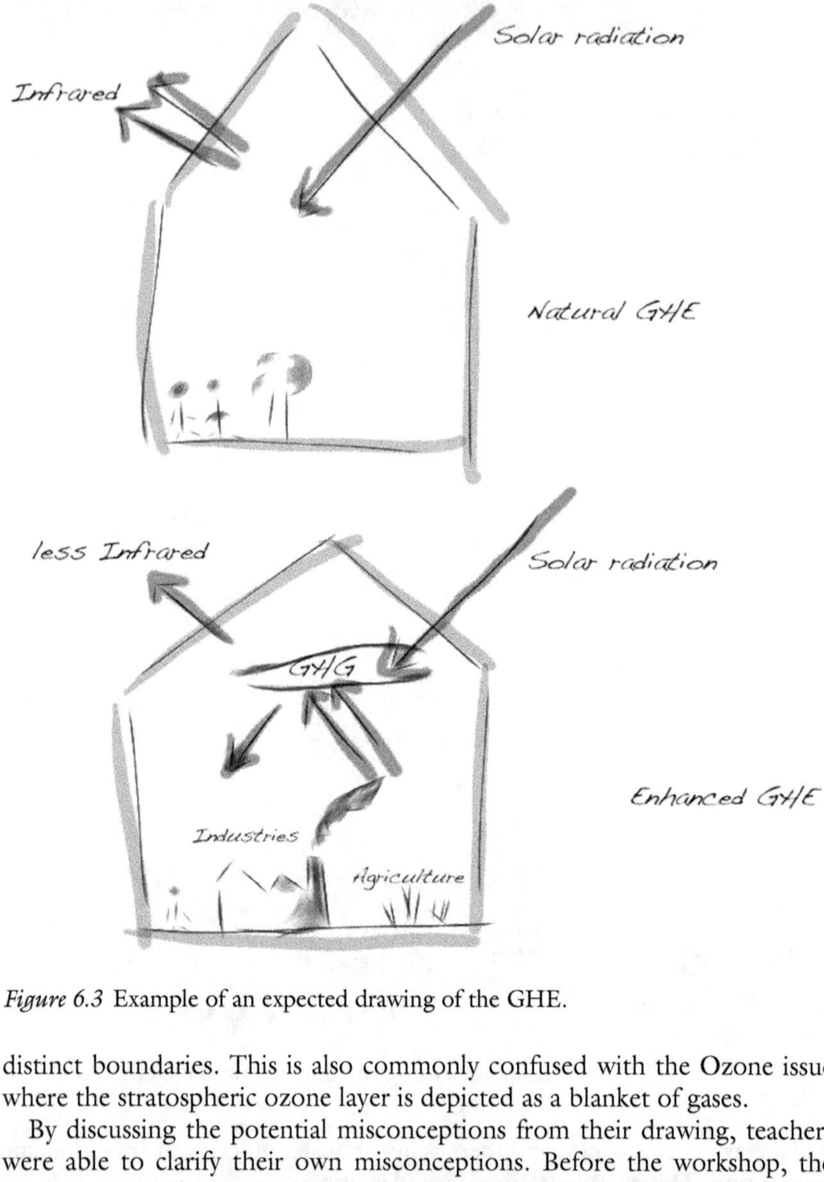

Figure 6.3 Example of an expected drawing of the GHE.

distinct boundaries. This is also commonly confused with the Ozone issue where the stratospheric ozone layer is depicted as a blanket of gases.

By discussing the potential misconceptions from their drawing, teachers were able to clarify their own misconceptions. Before the workshop, the teachers were asked to jot down what causes climate change, and they were asked to do the same after the workshop. Table 6.9 highlights the conceptual change in four of the teachers after the workshop.

It is interesting to note that although the teachers did not have many misconceptions in their initial answer to what causes climate change, the explanations became more nuanced after the workshop often citing opposing views to those they had before the workshop.

At the end of the workshop, the teachers developed a toolkit comprising the conceptual knowledge required for the three topics as well as notes on

Table 6.9 Conceptual changes before and after the PD workshop

Pre-workshop	*Post-workshop*	*What has changed?*
Human activities: globalisation, agriculture, urbanisation, transport, deforestation, industrialisation, greenhouse gases	Global warming isn't just caused by human activities	Awareness of other cause/s of climate change (such as global warming) though it is not explicitly stated what these other factors are
Latitude – change in temperature leads to change in pressure, which leads to pressure/wind belts. Shift in ITCZ, why Sahara desert is so dry? Distance from the sun. Angle of incidence due to Earth's tilt. At local scale – topography, relief, aspect, vegetation density, land surface	Climate is not just temperature and precipitation in a climograph. Climate and weather have the same variables	Cognisance of the different factors that make up climate
Natural causes, human activities	Cumulative effects of an interplay/interaction of natural and human causes	The interaction between natural and anthropogenic causes
Longitudinal location, physical landforms, proximity to large bodies of water: maritime influence. Human activities: global warming. Microclimates: cities	Climate changes due to solar and terrestrial effects. Broader categories (more encompassing)	Teacher elaborated more on the natural cause of climate change post-instruction. She/he did not mention about anthropogenic effects of climate change

the instruction approaches that were co-constructed by the group of 15 teachers in the workshop.

Beyond workshops and seminars – lesson study as a method

By referring to Chang, Pascua & Ess (2018), the example of this study not only provided the pedagogical approach of using refutation texts but also presented the use of the lesson study approach to teacher professional development. This takes our discussion of teacher education beyond just the traditional forms of professional development in seminars and workshops.

This study used lesson study (LS) (Guzzetti et al. 1997) as a technique to deploy refutation-based training in the tradition of school-based classroom interventions. LS, which has its origins in Japan, is the most frequent kind of ***konaikenshu***, or in-school professional development. LS is regarded as the "linchpin of the improvement process" (Stigler & Hiebert 2009, p. 111). LS, which is a direct translation of the Japanese word *jugyokenkyu* (*jugyo* for lesson; *kenkyu* for study or research), refers to the assessment of teachers' teaching techniques by other teachers, through a collaborative process of lesson discussions, planning and observations (Fernandez and Yoshida 2004). It is a problem-solving activity that entails iteratively developing, teaching, observing and reteaching lessons (Stigler & Hiebert 2009).

The LS cycle often involves small groups of teachers organising a research lesson that one delivers while other members of the team watch, with an emphasis on the learning and involvement of selected case pupils (Dudley 2013). The lesson is then analysed and improved in preparation for future teaching to additional groups (Hiebert & Stigler, 2009). The emphasis on observing students' learning throughout a class as the key to pedagogic improvement, rather than on an individual teacher's performance, separates LS from other kinds of professional learning (Cajkler et al., 2014). With such attention, instructors must shift and may be compelled to abandon assumptions about their own understanding of the topic.

According to the students' replies in Halvorsen & Lund's (2013) examination of LS as employed in history, the instructors' struggle on the first run to "abandon" what they taught was the essence of the lesson. As they identified flaws in the lesson design, they had to revisit the methodology and implement a revised lesson plan. The value of the lesson in assisting students to overcome their obstacles, rather than the performance of teachers in giving the lesson, shapes professional learning in LS (Lewis, 2009). The LS cycle also encouraged teachers to reconstruct students' thinking and arrange lessons based on their models of student thinking.

When utilised as a professional development tool, LS promotes teachers' reflective thinking about teaching, particularly when instructors work in a learning community. They not only prepare, observe and evaluate common lessons, but activities like pre-lesson planning and post-lesson evaluation sessions allow for extensive preparation and in-depth reflection on the quality of teaching and learning (Lewis, 2009).

The intervention in Chang, Pascua & Ess (2018) was a lesson study with the geography department's teachers. The study team met with two teachers to define the research aims and the expected learning results from the intervention as well as the use of refutation texts and the pedagogical approach to teaching the topic. The teachers and researchers designed the initial lesson design, which would be implemented using refutation text, during three rounds of discussions (discussed in Chapter 5). Several teachers from the geography and history departments as well as pre-service teachers on school attachments, were present for the first round of instruction. Their thoughts on the use of refutation text and instruction were solicited as soon as they watched the lessons. The input was then utilised to inform the refining of the lesson plan, which was completed by another teacher in the following round. Teachers' opinion was acquired through both a written evaluation and post-lesson interviews. Student learning artefacts were also gathered and labelled as research objects. Furthermore, the two instructors who used the pedagogical approach were interviewed.

The teachers who attended the sessions agreed that rebuttal texts help students confront misconceptions and reconsider preconceptions. In their reading, they become more aware and sensitive. According to one geography instructor who watched the courses, the text itself represents a "gradual release of responsibility to the students" by prompting them to be vigilant and responsive in their reading.

Refutation text, as a revision tool, requires content mastery. For a topic where misunderstandings are a barrier to learning correct concepts, the technique can assist teachers to build on existing literature and student suggestions to probe further (e.g., finding main causes), thus converting detected misconceptions into pedagogical resources. In refutation-based pedagogy, assessment is critical. Formative evaluation methods are important for ensuring that reformed knowledge is really understood. Scaffolding is required so that students may draw on knowledge or cross-reference sources to counter assertions. Detecting errors necessitates trust in one's expertise and comprehension.

Summary

This chapter reviewed the key practices in teacher education in EE and ESD as the basis to advance the framework for professional development in CCE. While EE and ESD target different cognitive and affective domains of educational objectives, CCE is concerned with the broad range of learning outcomes from cognitive to affective and even endeavouring to have students change behaviour that is pro-climate. Indeed, CCE *seeks to help learners develop knowledge, skills, values and action to engage and learn about the causes, impact and management of climate change*. Through the two teacher professional development workshops that identify the learning outcomes for CCE and propose a method of conceptual change to help students clarify misconceptions, a teacher is provided the opportunity to reflect on the questions of "what to teach" and "how to teach it" for CCE. While a meaningful curriculum and thoughtful design of instruction, coupled with teachers at a suitable readiness level are necessary conditions for a successful implementation of CCE, there remains a key area of concern – how do we know if the students have learnt. How do we know if students will do something about climate change? Assessment of and for CCE will be discussed in the following chapter (Chapter 7) to elucidate its relationship to curriculum and instruction for CCE.

7 Assessment in CCE

One fundamental issue about climate change education (CCE) is how to gauge a person's competency. Is it enough that a child goes through an assessment and passes the exam with flying colours? Is knowledge acquisition adequate? If we aim for attitude and behavioural adjustments, what are our indicators and measures? How do we build a citizenry that has the ability to discern, dissect and critique data about climate change? Good assessment practices in CCE can help teachers know how well they are teaching and how well the students are learning. If all the students in a programme do well, is it because the programme has been well delivered or is it because the material is too easy for them? It could also be that the students are extraordinarily capable and learn the material well despite the quality of the teaching. Conversely, the assessment provides teachers a profile of each student's learning, which will help the teacher adjust the instruction accordingly to the needs of the child. These could include the sequencing of topics, selection of content material to elaborate on, the repetition of material if needed, or even to decide if something need not be taught at all. Certainly, assessment serves more pragmatic reasons such as reporting students' progress to parents, school leaders and other stakeholders. This chapter will explore the issues with assessment for CCE with suggestions on how a good assessment for CCE can be designed.

What makes for good assessment in CCE?

Assessment is essential in teaching and learning because teachers require feedback on their practice as well as on the learning of their students (Voltz, Sims, & Nelson, 2010). Testing also serves a diagnostic purpose, helping teachers to determine different students' learning needs in order to better differentiate assessments for different student profiles. Finally, assessment serves as a mechanism to describe student progress (as well as instructor performance) to stakeholders such as school administrators and parents. Given these various functions, it is critical that the assessment used is reliable and valid. The results gathered for an individual should be similar even if it is administered at different times for the child and that it measures the outcomes that it is intended to measure. This is vital so that we can draw sound

DOI: 10.4324/9781003093800-7

conclusions about what the student has learnt. Consequently, good assessment tasks will not only help in determining how a student performs for the particular task, but it represents how well a student will perform on the learning objective for that task.

The claims or inferences taken from the test findings are concerned with validity of the assessment, whereas reliability in assessment relates to the consistent replication of results when using a test. In other words, validity is the accuracy with which assessment findings are used to represent a learner's performance or inherent traits. Criterion-related validity is concerned with test kinds and performance prediction accuracy, whereas content validity is concerned with an individual's "performance on a defined universe of tasks" (Shepard, 1993, p. 408). "[W]hen making inferences about unseen traits such as intelligence or anxiety" construct validity is required (Shepard, 1993, p. 409).

According to Stobart (2001), issues on assessment validity in the literature include whether the idea of validity should incorporate the topic of assessment consequences. Messick (1995), for example, contends that the validity of a test should reflect "potential and actual social consequences of applied testing" (p. 20) in addition to the content, criteria and construct validity that statisticians refer to. He contends that all evaluations have either good or negative societal implications. While standardised testing may promote student learning and motivation, as well as provide students with roughly equal access to classroom information, the disadvantages include simply teaching pupils to pass tests, rather than teaching them the skills they need for success in the world. There are also concerns over the impact that standardised testing has on teachers' motivation, with many teachers considering the impact of testing on their personal lives to be a significant detriment.

While content, construct and criterion-related validity are important considerations in test interpretation, the utilisation of test findings has social implications. Some researchers agree with this position (e.g., Crooks, Kane, & Cohen, 1996; Stobart, 2001), while others, such as Popham (1997) and Mehrens (1997), disagree, arguing that the social consequences of assessment are beyond the responsibility of test setters and should be separated from validity arguments.

In general, concerns about consequential validity focus on the impact of the assessment process on pupils. Crooks, Kane, and Cohen (1996) describe risks to consequential validity as the failure to attain intended results and the development of unwanted consequences. The biggest threats to consequential validity include low confidence in the assessment's appropriateness and fairness, as well as low student motivation. Stobart (2001) suggests that this necessitates a review of inappropriate standards as well as an evaluation of aspects of the assessment process itself, such as assessment conditions, nature of assessment tasks, weighting of different aspects of tasks and scoring criterion, based on Crooks, Kane, and Cohen (1996).

In thinking about consequential validity for CCE, we should begin with the question of what is considered core to the purpose of CCE. If the longer-term goal of CCE is to enable a more informed citizenry that is able to take positive environmental action that will mitigate climate change impact, then the way that assessment is designed must measure the range of cognitive, affective and behavioural learning outcomes that will lead to this goal. In essence, the framework presented in Chapter 2 outlines the kinds of knowledge, skills, attitudes, values and behaviour related to climate change causes, impact and management that we would like to see students learn. In some ways, this is similar to Delors' (1998) report to the United Nations Educational, Scientific, and Cultural Organization, which identified four pillars as fundamental principles for establishing education for the 21st century. Delors (1998, p. 97) advocates for an integrated approach to formal education based on four pillars of learning, which include:

1. learning to know – a wide general understanding as well as depth in a few subjects;
2. learning to do – to develop not just vocational competencies but also the skills and ability to deal with a variety of situations;
3. learning to be – to develop one's personality and to be able to act with increasing autonomy, judgement and personal responsibility;
4. learning to live together – through cultivating empathy for others and a respect for interdependence.

Griffin and Care (2015) argue that developmental taxonomies, such as Krathwohl's (2002) revision of Bloom's taxonomy of knowledge domains and cognitive processes, may be utilised to categorise and understand task requirements and student task responses. Indeed, Krathwohl's taxonomy indicates that while constructing assessments for CCE, it is necessary to incorporate both curriculum material and cognitive processes. Teachers will then set the criteria to be evaluated, allowing pupils to demonstrate their ability to achieve those goals (Cohen, 1995).

In considering how assessment could support CCE that is of social consequence – learning to know, do, be and live together – there has been an interesting development in Singapore over the last decade, where levels-grading and field-based geographical investigation have become a key component of the high school geography subject assessment. The Singapore example in geography education will show how, from the start, consequential validity may be a crucial element of curriculum design.

Incorporating consequential validity through changing the assessment type

Secondary school geography in Singapore has adopted a levels marking approach for part of the national written examination paper since 2007 (Singapore Examinations & Assessment Board, 2010). Since 2017, levels marking and questions focusing on field-based inquiry have been included in the national level examination. The fundamental motivation for the change was to allow examiners to get evidence of students' capacity to analyse and evaluate geographical problems in addition to simply reporting or explaining them. At the national examination for Geography, students were previously expected to complete 40 multiple-choice questions and four structured essay questions. Each relevant point in the structured essay questions was given a score.

The changes in 2014 required students to answer questions about how field-based geographical investigations can be undertaken; how data can be collected, organised, analysed and presented; and what conclusions they can draw based on their findings (Singapore Examinations & Assessment Board, 2014). While this example in geographical education shows that having assessment that is aligned to the goals of the curriculum to get students to investigate a geographical phenomenon – to have the geographical thinking, this is perhaps still not enough for CCE. There is also a need to include behavioural and attitudinal aspects in assessments intentionally, and if the intent of CCE is to see behavioural or mindset change, then we need to encourage independent and critical thinking (Jickling, 1992; Aldrich-Moodie & Kwong, 1997).

Some argue that advocating for values education and social change for the environment (as suggested by Fien, 1993; Huckle, 1985; Morgan, 2012) is anti-educational (Williams, 2008), but not allowing for any examination of behavioural and attitude learning results would be worse. For example, some studies have discovered that teachers are uncomfortable and hesitant to advocate for the environment in their lessons (Tomlins & Froud, 1994; Cross, 1998), but are more at ease presenting a neutral position while discussing a variety of different viewpoints on environmental topics (Tomlins & Froud, 1994; Cross, 1998, Cotton, 2006). Some instructors, on the other hand, believe they are accountable for fostering environmental attitudes and behaviours (Grace & Sharp, 2000). In fact, Ho and Seow (2015) discovered that teachers in Singapore interpret and perform the identical geography curriculum differently depending on their ideas about the purpose of geographical education.

Good assessment should have the content and scope of the assessment aligned to the learning outcomes for the curriculum (standards or syllabus). This is critical in helping educators design assessments that will help improve student learning that will take them beyond "learning to know" to "learning to do, be and live with each other". In the case of CCE, the framework to describe the domain of curricular content knowledge will be useful in determining the scope of assessment tasks. Indeed, we need to know if students have *developed knowledge, skills, values and action to engage and learn about the causes, impact and management of climate change.*

The modality of assessment has also been discoursed within the dichotomy of summative and formative assessment. If summative assessment is used as an accountability measure to ascertain how much individuals have learnt, then formative assessment integrated with instruction has the potential to increase learning (Hagstrom, 2006). Indeed, classroom assessment has often been considered weak (Black, 2000) and there is a desire to improve classroom assessment that can truly help students learn better.

However, the prevalent sort of assessment in many schools is Assessment of Learning. Its object is to ensure learning and determine learners' advancement in school. It generally provides an indication of the relative position of an individual within a cohort of students. This form of assessment is normally at the end of a learning activity, such as the end of a topic or a project, usually manifested as tests or exams that material from the content taught.

Assessment for Learning, as its name implies, shifts the focus from summative to formative assessment, from determining normative performance to providing descriptions that potentially offer clues as to how the activity can be modified to support the needs of the student's learning. Normally, a wide range of data is collected with the end of improving student learning. Tasks are designed to uncover what students know or can do and are subsequently used to improve the process. Grading is not meant to make comparative judgements about where students stand in a class but rather to help the teacher in determining each learner's strengths and weaknesses and offer them feedback that will advance their learning.

Grant Wiggins and Jay McTighe propose that curriculum should be designed backwards from the assessment tasks (Wiggins & McTighe, 2005). In the context of CCE, it is first important to decide what students should be able to demonstrate in terms of their knowledge and skills before we decide on what aspect of climate change, we want to teach them. While summative assessment tasks seek to determine a single score for the purposes of establishing a general level of progress for each child, assessment tasks designed to help teachers and students improve individual learning needs to have specificity in the criteria used for assessment so that we do not just determine how well a student has performed, but also why they have performed this way (McTighe & Self, 2002). Indeed, the three elements of an effective performance assessment would include a meaningful context, the thinking process and an appropriate product or performance (Wiggins & McTighe, 2005).

Good performance assessment tends to be authentic and situated within some meaningful context rather than having a student answer questions randomly across topics in a traditional test. A good performance assessment should also clarify the thinking process of the learner and provide an indication of the ability to apply the knowledge to a new situation. Further, there is a need to avoid tasks that do not relate to the content of what is being assessed (Wiggins & McTighe, 2005). In other words, the authors reiterate that the assessment needs to be valid.

Backward design to support assessment for learning CCE

It is important for teachers to take an assessment perspective right from the beginning if they want to develop meaningful performance assessment tasks that will indicate what students have learnt. In other words, in answering the question of how you know what your learners have learnt, one will seek to ask what would count as evidence of the learning (Wiggins & McTighe, 2005).

Rather than asking what would a good learning activity look like, it would be important to first determine the criteria with which to determine what counts as evidence that the child has learnt something in an assessment task (McTighe & Self, 2002). Indeed, performance criteria outline what tasks are expected of students and how each component will be evaluated (Wiggins & McTighe, 2005).

> Creating effective assessment tasks requires thinking through curriculum content to establish learning outcomes, then designing performance activities that will allow students to demonstrate their achievement of those outcomes and specifying criteria by which they will be evaluated.
>
> (Cohen, 1995 p. 1)

This backward design approach whereby the assessment is first determined before designing instruction "promotes reflection on all the larger issues of teaching" (Cohen, 1995). In addition, by engaging their peers in a review process whereby "teachers compare one another's assessments with content and assessment standards" (Cohen, 1995), the task design can be refined and consequently result in a better design of instruction. In the first step to the backward design, the goals of CCE must first be determined before meaningful tasks that are situated in a real-world context can be designed.

Assessment of and for CCE – what are the goals?

In referring back to Figure 2.4, the knowledge domains of learning outcomes for CCE can be described as

I. Climate change causes
II. Climate change impact
III. Climate change management

The cognitive and affective domains of learning outcomes for CCE can be described as

1. Knowledge
    a. Recall
    b. Understand
2. Skills
    a. Analyse
    b. Evaluate
3. Values
4. Action

In referring to Table 2.3 which shows exemplars of the outcomes of CCE, one can determine the learning outcomes with specificity. Specifically, we need to know which aspect of helping students ***develop knowledge, skills, values and action to engage and learn about the causes, impact and management of climate change*** we want to measure.

After determining exactly what goal it is a particular unit of instruction endeavours to achieve, it is important to decide what would count as evidence. For instance, if the outcome of a unit of instruction is for students to be able to use multiple sources and argue for or against human-induced climate change, then the evidence of learning would be for the child to demonstrate through a task to argue, substantiate and evaluate for his or her stand on the issue. Similarly, if the unit of instruction seeks to change the child's personal habits to reduce his or her carbon footprint, then the evidence of learning must be measured against the actual action the child has taken. Inherent in this exemplification of determining the evidence of learning is that the nature of the task and the context for the task design should take into consideration the following:

1. Does the task truly match the outcome(s) you're trying to measure?
2. Does the task require the students to use critical thinking skills?
3. Does the assessment use engaging tasks from the "real world?
4. Can the task be used to measure several outcomes at once?
5. Is the task feasible?
6. Is the task clearly defined?
7. Are instructions for teachers and students clear? Does the student know exactly what is expected? (Chicago Public School)

A Goal-Role-Audience-Situation-Product/Performance (GRASP) format of determining the goal, role, audience, situation, purpose and standards can be used in the design of a performance task. The following example of a performance task to assess if students are able to use multiple sources and argue for or against human-induced climate change is used as an illustration of this format.

> Goal:
> Your goal is to comment on a blog post and share your opinion on whether the climate change that we are experiencing is caused by humans.
>
> Role:
> You are an active participant on many local blog sites that include topics on the environment and society. You are also the secretary of your school's green club and you have written several school essays on the topic of climate change as this is your area of interest. Naturally, you have done a fair bit of research on the topic.
>
> Audience:
> The audience is the readers of the blog page of "Mr Green".

Situation:
The blogger who comes under the screen name of Mr Green has recently been ranting about how climate change is all due to human activity. Having done a fair bit of research on the topic, you have decided to post a comment to balance the perspectives and then share your stand on the matter.

Product Performance and Purpose:
You may choose a stand that is either for or against the argument that climate change is human-induced. Be sure to include examples and reasons for your arguments and also indicate how you have come to a conclusion on your stand.

Standards and Criteria for Success:
Your comment needs to offer perspectives for or against the claim that climate change is human-induced, provide substantiation using examples and offers a stand.

In the example above, you will observe that the key criteria used in determining the success of the task are to argue for or against the stand, substantiate with examples and adopt a stand. These are clearly targeted at the skills of analysis and evaluation. By adopting a stand, the student is also afforded the opportunity to inform his belief about the issue.

Inherent in the design of the performance task will be the need to design instruction to support this task. To this end, the instruction must necessarily determine if the child has the prerequisite skills to argue for and substantiate a claim. If this is not present, the learning activities will have to include this. Similarly, it assumes a certain level of writing skills for this activity. The subject matter knowledge also needs to be accessible to the child. Whether the information is taught or an inquiry process involving some web research is needed, the subject matter for this task must be carefully and purposefully included within the curriculum time leading to this task.

There is also a necessity to set the child up for success. By providing a well-designed set of evaluation rubrics, the student can also assess himself/herself along the way thereby allowing the student to use the assessment as a learning process – an assessment for learning.

This approach to designing assessment for CCE is primarily relevant for formative assessment for learning. In the case of summative assessment, the specific learning objectives are often prescribed by a national examination board or the school. Commonly, the learning objectives are in the cognitive domain and do not address the values and action parts of CCE. In fact, there is a proliferation of focus on the knowledge about climate change in the formal curriculum. However, there is a need to balance both the learning outcomes for summative and formative assessments. As one teacher rightly pointed out, test scores are not reliable indicators of whether a person cares for and acts for the common good of other persons and of the Earth.

Assessment of and for CCE in Singapore

In terms of summative assessment, the 2007 upper secondary Geography curriculum (Singapore School Examinations & Assessment Board, 2010) adopted a new "levels marking" approach for part of the national written examination paper. The aim of this initiative was to measure students' ability to discuss and evaluate geographical problems. Consequently, some teachers voiced apprehension over the deviation from a common practice of the "point-marking" approach and felt that perhaps they needed training to be prepared for this new approach. Indeed, the introduction of levels marking was intended to prepare students "to meet the challenges of an increasingly globalised world … to promote critical and creative thinking skills, and to nurture problem-solving and independent learning abilities in students" (Sellan, Chong, & Tay, 2006). Up till this point in time, the national examination for Geography has comprised two examination papers, one was a 40-question multiple-choice papers and the second paper included a dozen structured essay questions, from which four questions are to be answered. A point-marking scheme in which marks were awarded for relevant and accurate points raised was used for the structured essay questions. In the new syllabus, the students answer two out of four essay questions for each of the physical geography and human geography paper. However, for each question, the last eight marks are allocated to a sub-question, testing application, comparison and evaluation, graded using the levels marking approach described above.

As Geography as a subject is intended to prepare the child to engage in the global issues of our time, in accordance with the *charter*, the development of the assessment format to test higher order skills supports this intention. Students have to evaluate the issue and then construct an opinion, after analysing the information provided. Nevertheless, the modality in the summative assessment of CCE is still unable to determine the types of values learnt and if it results in behavioural change.

In terms of formative assessments in CCE in Singapore, these occur in the form of class quizzes, reflection papers or project works. As to the extent of learning, teachers claim that there is little evidence that students imbibe the skills and concepts taught on climate change well. Similarly, students are not confident of the knowledge they hold about the topic and do not see much relevance of it in their daily lives apart from scoring well in the examinations.

As a follow-up from the baseline understanding of what students know about climate change, I have also conducted a few in-depth interviews to understand other aspects of their climate change learning. In the interviews with students on their perspective on CCE, the range and extent of assessments vary in the students' responses. Three of those interviewed claimed that they do not remember having gone through any form of assessment while the rest reported having quizzes, writing reflection papers and doing project works. The questions asked often revolve around the topics of natural vegetation with regard to global warming and greenhouse gases, the cause, effects and mitigation strategies for climate change. Others remember interpreting graphs and statistical data such as mean annual temperature.

In terms of project work, some of the students remember doing a poster, others a representation of how recycling is beneficial to the Earth. They were then asked to do presentations based on their output. Others do not remember being engaged in any project work related to climate change.

In the teacher focus group discussions, they expressed their reservations regarding the efficacy of assessments to truly measure students' understanding and capability to tackle the challenges posed by a changing climate based on what they have learnt in in-school instruction. At the end of the day, one has to look at whether this was translated into action. The teachers observed, however, that the learning outcomes fail to indicate whether students have truly learnt the skills and concepts of CCE. Evidently, test scores are not reliable indicators of whether a person cares for and acts for the common good of other persons and of the Earth. As one teacher stated, "Action is different from tests and scores, and action is very difficult to measure actually". Then again, another lamented that even as she tried hard to connect concepts with action and students may actively participate in it, and at the end of the day, the learning may not be that much and is simply "*thrown out of the window*". As one student mused,

> Actually, I think exams are very effective of making us remember things. But erm…but (if) you remember, do you do it? I don't, you know. I remember, I know everything. I know things that I'm supposed to remember, but I don't…(long pause). It's empty, lah!

Assessment aside, the researchers probed into students' top-of-mind recollections of climate change concepts. It appears that what students remember from classroom discussions are the consequences of the phenomenon. Such include the concepts of global warming, pollution, the greenhouse effect, sea level rise and floods. They also asserted that they remember the case of chlorofluorocarbons while some specified learning about international treatises like the Kyoto Protocol and the Montreal Protocol.

Mitigation strategies were also high on the list of what students remember, most prominently the 3Rs, saving electricity and use of public transport in place of private cars. The interviewer generally noted that students did not have much to say when asked the question of what they remember most from their lessons on climate change. They took some moments to ruminate and at the end of a long pause, did not supply much information. One student stated the frame of mind that the other interviewees were not able, or were not comfortable to articulate:

> I don't think most people would bring home what they actually discussed about. And some people would forget about it. Some people actually take down notes to study for the exams. I think after the exams, everybody would just (pause) yeah, forget about it.

The need to incorporate climate change topics at different levels, not just Secondary 2 or 3 was also given emphasis. One student aptly illustrated this point:

> We have to have to have a mix of everything. If not, it's just like...(if) only for that sole year, I just focus on climate change and then forget about it for the rest of my life.

While enhancing knowledge and improving understanding of the climate change topic seems to be the key educational strategy for Singapore students and students in other countries alike, formal curricula that systematically educate youths on the issue have not produced many results over the past few decades.

In an attempt to reverse this trend and generate practice that encourages assessment for learning in CCE, some of the outcomes from the professional development workshops were assessment ideas that teachers developed during the programme. As the copyright of these artefacts belongs to the teachers, I am unable to share them with the readers. However, I shall abstract the discussions and provide an example of a performance task for a Geography lesson.

Again a GRASP format of determining the goal, role, audience, situation, purpose and standards was used in the design of this performance task. The following example of a performance task to assess if students are able to understand the rationale behind setting the air-conditioner thermostat to 25°C and take action on it.

> Goal:
> Your goal is to convince fellow students in Secondary 3 that it is possible to reduce your carbon footprint through the way you use your electricity at home. You will need to use specific examples to convince them.
>
> Role:
> Being the chairperson of your school's environmental club, you have been asked by your teacher to give a half-hour presentation on encouraging your fellow students at assembly time during Earth Week. Knowing that you have been an activist and that you practice what you preach, the teacher has chosen you to inspire fellow students into taking action to manage the climate change issue at a personal level.
>
> Audience:
> The audience is the cohort of Secondary 3 students at your school who, in your observation, seem to be oblivious to their responsibility to the environment.
>
> Product Performance and Purpose:
> Your presentation should be backed up by good research that explains how the mitigation strategy at a personal level works. In particular, you

will want to explain how setting the thermostat of the air conditioner to 25°C will have an observable reduction in your carbon footprint. While you cannot measure the actual amount of carbon emissions you can reduce, you may use actual data from your daily life, through your own pro-environmental practice to illustrate the point.

Standards and Criteria for Success:
You do not have to make all the students believe in you. You will however need to offer compelling evidence to convince them. Perhaps you should be able to convince them that action speaks louder than words and that you could show them that you do practice what you preach.

The task above will require the student not just to demonstrate an understanding of how a specific mitigation strategy can work, he or she has to evaluate its effectiveness through the personal action that has been suggested. Of course, the true test of whether this will result in any behavioural change can only be measured some time down the road. The design of this task, however, at least provides the opportunity for the child to examine his or her own action in perspective.

Summary

By examining the key principles of assessment practices and by advancing a method of performance assessment that is based on authentic tasks such as activities, or problems that require students to show what they can do, it ties back to the outcomes for the curriculum and the instruction for CCE. Indeed, for CCE to be successfully implemented, all three components of curriculum, instruction and assessment must be designed synergistically, set against the framework of learning outcomes described in this book. However, there is still a vexing problem whereby there is a reported gap between student's attitudes and actions.

In a comparative analysis across cases in terms of country, curricula implementation, the grade level of study and the time at which the studies were conducted, there has been little progress in helping students correct these misconceptions about climate change over the years (Jeffries, Stanisstreet, & Boyes, 2001). How such unconventional ideas are generated and become commonly held beliefs concerns some researchers. Students' ideas are formed from common daily experiences, including opinions formed by the media (Svihla & Linn, 2012). Indeed mass and social media provide the bulk of informal learning among students with the mass media as the most prevalent resource (Hansen, 2010). Specifically, the limited time devoted to discussing environmental issues in the classroom has caused students to turn to alternative sources of information, worsening the problem as the information gathered is coloured the way information tends to be sensationalised.

While the child is in school for about 6–8 hours each day, the other two-thirds of his or her day are spent outside a formal education context. Indeed, the influence of informal education cannot be ignored in understanding how CCE can effectively transform our young generation in acting for the common environmental future of the Earth.

8 Lessons from CCE for public education

While formal education occupies a significant bulk of the quantity of learning students are exposed to on climate change, in Singapore as elsewhere, it is a fact that the classroom does not figure as the top source of students' environmental knowledge. Tan, Lee and Goh's (1998) study for instance provides evidence that young Singaporeans learn more from alternative sources, mainly from print and broadcast media about environment-related concepts and events. Additionally, a small percentage indicates that they engage in co-curricular environmental activities such as special courses, talks and exhibitions. How do we build a citizenry that has the ability to discern, dissect and critique data about climate change? The role of public education should also be considered as it has implications on how a student learns outside of the formal classroom context, as well as for life-long learning opportunities for learners who have graduated from formal education programmes. A specific study on grassroots action will be included in this chapter to illustrate these issues.

Informal CCE in Singapore

Against the backdrop of numerous activities aimed at raising public understanding and awareness of climate change, the question of how Singaporeans can be prepared for climate change through education remains high on the agenda of national agencies such as the NCCS. In an attempt by the Singaporean government to test the social temperature of its citizens regarding climate change issues, a survey conducted by the National Climate Change Secretariat (NCCS) in 2011 found that Singaporeans had awareness and expressed public support for projects to address the climate change issue and its threats. Among the breakthrough suggestions put forward by the respondents is the inclusion of climate change topics in the formal school curriculum (National Climate Change Secretariat, 2011). While the study did not elaborate on how heightened awareness translates into action or attitude change, it did provide an important vantage point from which we can discuss the state of informal climate change education (CCE) in Singapore. In line with the Singapore government's explication that there should be a school-civil society-business-government collaboration in instilling environmental consciousness among young minds (Mekani & Stengel as

DOI: 10.4324/9781003093800-8

cited in Tan, Lee, & Goh, 1998), we can explore the extent to which activities comprising this kind of informal curriculum open up opportunities for CCE in Singapore and possibly extending the findings to other contexts in the world.

The case study of Singapore in this book confirms the claims of the literature that there is a prevailing lack of understanding of climate change – its causes, impacts and measures to alleviate its effects among students. In essence, the findings reflect Koulaidis and Christidou's (1999) observation that climate change integration into the formal curriculum is patchy and sporadic. School children, while aware of common concepts related to climate change, do not possess accurate knowledge about it. In fact, they hold unorthodox explanations about the science behind the phenomenon. For instance, they neither understand the enhanced greenhouse effect nor the natural causes of climate change. Even when CCE is incorporated formally into school curriculum like in the case of Singapore, children's exposure to alternative sources of information is so great that there is no escape from informal and public climate change discourse (Hansen, 2010).

Environmental initiatives from various social sectors in Singapore (business, civil society) provide alternative routes for learning. Activities are being organised on an ad hoc basis targeting Singaporean youth in particular and the general public. The bulk of these activities are awareness campaigns to move up public responsiveness. In order to understand the state of Singapore's informal curriculum on CCE, the findings are discussion and organised according to activities by public organisations and private enterprises and by the nature of activities.

In terms of education, Singapore has recently introduced the Eco Stewardship programmes to complement an extensive set of environmental campaigns in the form of public lectures, forums and exhibitions hosted by both government and non-government organisations. Apart from formal classroom learning, some schools have co-curricular activities centred on environmental awareness and conservation in the form of informal education. Environmental education is incorporated into other subjects such as science, social studies, geography, civics and moral education (Kong, Yuen, Sodhi, & Briffett, 2000). However, schools and special environmental classes are not perceived by students to be important sources of environmental knowledge (Tan, Lee, & Goh, 1998). Considering that the youth is plugged into online media most of the time, Chib, Chiew, Kumar, Lim, and Ale (2009) suggested that the Internet is a viable medium to reach the youth sector of Singapore in affecting their environmental attitude.

Issues of climate change have become more prominent in Singapore in the last couple of years and because the leadership in government is discussing them in the public arena, the momentum in focusing and addressing them will only accelerate in the future. This recognition of climate change and environmental impact of various kinds due to human action will influence schools to increase their students' focus on them. Already there are some promising signs that this is happening. There are more frequent and positive reports noting that primary school children were now more involved with environmental issues and studying topics such as climate change and renewable energy (Gilani Research Foundation, 2009).

Whether people will have improved understanding of climate change after exposure to different information sources was what concerned Reynolds, Bostrom, Read, and Morgan (2010) in their study of laypeople's mental models of climate change in Pennsylvania. Comparing their 1991 and 2009 samples, the authors conclude that 17 years of debate did not advance understanding significantly. While the latter respondents display improved awareness and comprehension as well as desirable behaviours, incorrect beliefs still persist. For example, the respondents fail to indicate awareness of global warming as mainly due to increased carbon dioxide in the atmosphere and that the main source of CO_2 is the combustion of fossil fuels. Other misconceptions noted are that climate change is caused by lunar cycles and that global warming causes skin cancer. The authors also noted that laypeople are not familiar with terms and questions posted to them, further arguing that this only highlights knowledge deficiency as these are only basic science concepts related to climate change.

Informal opportunities for learning about climate change are made available through initiatives by government agencies, the business and civil society sectors, interest groups and community councils. Projects initiated by these bodies often take on an environmental slant with a focus on individuals coming together to collectively contribute to reversing the negative effects of climate change. These projects are typically co-designed and implemented with schools to target the young population of Singapore.

Alternative modes of learning through the informal curriculum will be also presented as available avenue for students to learn. These are initiatives by certain government agencies, the civil society sector and partnerships between schools and members of the business communities and local councils.

Thus, the informal curriculum presents a fertile ground to initiate advancement in learning. However, the known weaknesses of informal learning are the lack of a defined structure and an overarching goal to link all initiatives together. The framework introduced in the earlier chapters will move public education in an informal setting forward. It is therefore paramount that CCE is understood by both the schools and partner agencies. The main benefit seen with this arrangement is that CCE will be incorporated into mainstream thinking, not as an isolated case of an environmental issue or a topic in class to be assessed, but as a prevailing discourse that everyone has a stake on.

CCE efforts by public institutions

Government agencies

As a leading organisation responsible for improving and sustaining a green environment in Singapore, the National Environment Agency (NEA) has liaised with different sectors and launched a series of educational programs for the public. The NEA declared 2018 as the Year of Climate Action. In another example, the program entitled "Everyday Superhero" (National Environment Agency, 2006) is an exhibit that encouraged the public to change their everyday habits and behaviours to reduce energy usage and improve fuel efficiency. Similarly, in April 2008, the NEA introduced the "10% Energy Challenge" to raise awareness of energy conservation and how to adopt an energy-efficient lifestyle as part of their daily lives. Singapore residents were challenged to reduce their home energy consumption by 10% between May and August 2008 and win prizes in a national lucky draw, the top prize being a fuel-efficient hybrid car. Unfortunately, Singapore's CO_2 emissions from fuel combustion are contributed in large part by electricity generation. The projected increase in emissions is from 38 to 71 million tonnes of CO_2 between 2005 and 2030 (Asia Pacific Energy Research Center, 2009).

Accordingly, the NEA hoped to empower Singapore residents to save energy and contribute to a collective effort to mitigate climate change. Apart from engaging the public in the battle against global warming, in order to enhance young people's understanding of climate science and raise their environmental awareness, the Clean and Green Singapore Schools' Carnivals organised by the NEA provide a platform for the schools and corporate partners to showcase their collaborative projects on climate change (Clean & Green Singapore, 2010). For instance, a school partnered with a battery company set up a solar-powered charging station for rechargeable batteries, which may facilitate students to better understand the benefits of solar energy to the environment and its applications. With regard to promoting CCE at the national level, NEA along with the Ministry of Education launched the "Youth for the Environment Day Program" in 2010 so as to foster young people to champion environmental ownership and renew their commitment to care for the environment (National Environment Agency, 2010). This program suggests that all the young people in Singapore should safeguard their living environment, nurture an eco-friendly and socially gracious culture and cherish the clean and green environment for present and future generations.

Riding on NEA's campaign, localised intervention strategies are spearheaded by Community Development Councils (CDC). For example, the Southwest CDC specifically targets households to educate and promote appropriate behaviour. It organised a group called the Eco Energy Auditors to advise participating homes on energy-efficient habits and to earn incentives for total energy saved. Another initiative, the *100,000 Native plants @Southwest*, was started to reinstate the natural heritage of Singapore by planting native plants in the Southwest District and secondly, to reduce the effects of global warming and climate change. Said CDC also embarked on an educational campaign on energy-saving habits like 25°C Campaign @Southwest (25°C being the recommended optimum air temperature) to inform constituents on how they can contribute to combating climate change through minor alterations in their urban lifestyles (South West Community Development Council, 2007).

An initiative called "Project E.A.R.T.H." (Every Act of Recycling Trash Helps) was organised by the NEA and Central Singapore CDC in 2009, which attempts to galvanise citizens from the public and private sectors to meet a "district goal of 10,000 tons of recyclables" to encourage recycling practices in the Central Singapore district (CSRSingapore, 2009). Since its launch, the "Project E.A.R.T.H." has managed to engage more people from residents, students and corporate partners in recycling activities, which in turn "fosters community bonding" (CSRSingapore, 2009). Efforts carried out by schools such as in-school recycling programs and outreach to residents (Recycling collection drive). To date, 23 constituencies, 22 schools and 14 private companies in Central Singapore District have participated in this initiative. In order to bring the initiative to another level, the NEA Central Regional Office and Central Singapore CDC have recently launched the "10% Total Project E.A.R.T.H. Challenge" with the purposes of attaining a 10% increase in (a) tonnage of recyclables collected, (b) the number of recycling activities implemented, (c) the number of schools engaged and (d) the number of corporate partners involved (CSRSingapore, 2009).

To prepare for the challenges of climate change and secure a sustainable living environment, the NCCS Singapore was established in 2010. The NCCS has organised a national climate change competition annually targeting tertiary institutions, junior colleges, centralised institutes, institutes of technical education and secondary or primary schools. This competition aims at promoting learning about climate change and nurturing concern and appreciation for the environment. Students are able to identify and propose solutions to address climate change issues by participating in the competition. It will also encourage students to innovate and develop new, original and viable solutions for climate change.

Informal settings like Singapore Science Centre also play an important role in educating the public about climate change. With the help from NEA and Shell Companies in Singapore, they have organised a climate change exhibition which showcased local content and a collection of exhibits from science institutions all over the world, such as the "Marian Koshland Science Museum in Washington DC, the National Museum of Emerging Science and Innovation in Japan, the Interactive Institute in Sweden, as well as Science North in Canada" (National Environment Agency, 2012). The exhibition employed interactivity to engage the visitors to think about the issue, achieve a deeper understanding and hopefully to change their personal habits both at work and at home to conserve energy and hence mitigate emissions of greenhouse gases. In addition, Singapore Science Centre also jointly organised a public forum on climate change and biodiversity with the Singapore Institute of Biology. This forum offers the public a platform to discuss about the biodiversity impact that simulates various aspects of climate change with scientists (Singapore Institute of Biology, 2008).

Partnerships

Collaborations across the sectors are consistently observed. One noteworthy program is the NEA's partnership with schools across the island in building up a network of student environmental advocates. To date, the NEA has trained some 470 students from 82 primary and 58 secondary schools. A good number of them have begun to give talks in their campuses encouraging their peers to adopt good environmental practices such as recycling. This program is complemented by the agency's launching of "Youth for the Environment Day". Together, these efforts constitute a key platform for young Singaporeans to advocate ecological ownership and be acquainted with proper values in caring for the environment. Recently, the program expanded to include pre-school students also as Little Green Ambassadors.

NEA also partners with schools in organising an annual project called Clean & Green Singapore Schools' Carnival. The Carnival is a national-level site for academic and corporate partners to showcase their collaborative projects under the Corporate and School Partnership (CASP) programme. Under this scheme, a National Weather Study worth S$1 million was implemented and sponsored in 2005 by a power generation company Senoko with support from the Ministry of Education and the NEA. The aim of the project was to encourage schoolchildren to study first-hand the effects of climate change with the help of a sophisticated weather-monitoring equipment installed in their school backyards.

Competitions

Competitions appear to be a popular method to engage the youth in meaningful discourse on the issue of climate change. Common in these contests is the aim to promote learning about climate change and to stir critical and analytical thinking among participating students. The NCCS, a government body, organises an annual project called the National Climate Change competition involving students from all levels of the education system. On the same note, Sembawang Shipyard Private Ltd., a shipping company, also held an annual competition called the "Green Wave" Environmental Care Project for schools in 2003. This is an initiative that promoted the rethinking of one's behaviour vis-à-vis scenarios brought about by a changing climate system.

Non-governmental organisations

A non-government organisation (NGO), the Singapore Environment Institute (SEI), started a program dubbed Programme for Environmental Experiential Learning (PEEL). PEEL provides hands-on learning geared towards expanding participants' knowledge about green issues management through a series of field visits to relevant sites around Singapore. Each PEEL trail provides participants with behind-the-scenes perspectives of the various environmental management strategies in Singapore.

NGOs such as the Singapore Environment Council (SEC) runs the School Green Awards (SGA). This is a voluntary environmental programme for students to showcase their environmental efforts so that students and teachers can become responsible advocates of the environment. (Singapore Environment Council, 2022). Schools with innovative green programs and individuals who have made an impact on environmental issues will be given recognition for their effort to preserve the environment. In addition, they set up the Tertiary Institutions Council for the Environment (TICE), to allow institutions to collaborate and compete in environmental projects, so as to strengthen environmental awareness and imbue stronger responsibility among tertiary students. To date, TICE has organised events, campaigns and competitions and has also provided a platform for institutions to share information on their projects and best practices. For example, environment scientists are invited to speak at TICE's activities and exchange views on the causes and impacts of climate change with students. For example, geography academics at the National Institute of Education were invited to give a talk at TICE Eco Camp 2009 on the topic of climate change and its impacts on Singapore.

In engaging the youth in environmental issues, the National Youth for Achievement Award (NYAA) Council has organised activities such as the National Youth Environment Conference. This conference provides an avenue for the participants to engage in dialogue over a wide range of environmental issues. The most recent conference was organised in March 2022. Other activity organised by NYAA Council includes the Active, Beautiful and Clean waterways Photography Workshop that supports young photographers enhance their skill as well as raise environmental awareness through photography workshops.

Lastly, educational organisation such as the Southeast Asia Geography Association (SEAGA) frequently organises educational forums to bring together different groups of people like educators, policymakers to discuss environmental issues. For example, in February 2009, SEAGA organised an international forum on climate change to bring in scholars, policymakers, entrepreneurs and concerned citizens to debate state and society responses to climate change and its implications, particularly for populations living in cities. In subsequent years, Singapore educators have taken part in several SEAGA conferences organised in Siem Reap and Jakarta with sustainability as part of the conference themes, in 2014 and 2017, respectively.

Local educational institutions

Compared to the actions taken by the governmental and non-governmental organisations as discussed above, local educational institutions have also stepped up their efforts in educating about climate change. At the National University of Singapore (NUS), students from the interest group, that is, Students' Against the Violation of the Environment (SAVE) have organised a NUS Fight Climate Change Green Carnival since 2006. This carnival serves as a platform to raise environmental awareness in the NUS community and further guide them to take action for the future of the environment. In another local university, Nanyang Technological University (NTU), the student group Earthlink NTU collaborated with Orchard Road Business Association and became a part of Celebrate Earth Hour event in 2011, which was to support and pledge to save the Earth through people's daily efforts in the process spreading awareness and inculcating a sense of belonging to the environment amongst students and the general public. In addition, NTU has introduced a compulsory course on Sustainability Society, Economy and Environment (called an interdisciplinary collaborative core course) for all undergraduates at the university in 2021.

Activities are not confined to tertiary institutions like universities; secondary schools also have organised various activities to engage students in sustaining the environment so as to tackle the impacts of climate change. For example, Yishun Secondary School organised educational trips for students to the Tzu Chi Foundation's educational recycling station (Ong, 2011). Prior to the trip, volunteers give a short presentation highlighting recent natural disasters to the participants in hopes of raising awareness of the issues of climate change and the importance of environmental protection. Then, during the visit, the students were encouraged to sort out recyclable items. As reported by an accompanying teacher, the students began to compost food waste collected from the school canteen to make fertilisers when they went back to school. It is apparently noticed that the educational visit to the recycling station has great potential to facilitate students to their learnt knowledge into actions.

Secondary schools, such as the Yishun Secondary School, also engage in organised educational trips for students to the Tzu Chi Foundation's educational recycling station. Prior to the trip, a short talk was given by volunteers citing recent natural disasters to impress upon the students the problems of climate change and the importance of environmental protection.

Private organisations

The private organisations have made a great contribution to the promotion of CCE in Singapore. By collaborating with schools and governmental and non-governmental organisations, these private sectors engage the community in learning about and for the environment that we live in. This section describes the actions that private organisations have taken to support the implementation of CCE in informal contexts.

As far as climate change awareness is concerned, a national weather study project to expose thousands of school children to study first-hand, the effects of climate change with the help of sophisticated weather-monitoring equipment installed in their own school backyards was held in 2005. This S$1 million project was conceived and sponsored by SENOKO, a power generation company, with support from the Ministry of Education and the National Environmental Agency (Lim, 2011). SENOKO has also been organising the

National Weather Study Project Competition since 2006, which has garnered the focus of schools and heightened the awareness of the impact of climate change on the environment (SENOKO, 2009). SENOKO also organised the Senoko Sustainability Challenge (SSChallenge) in 2013 with a repeat run in 2016, an outreach programme aimed at reframing climate change in the Singapore context. Sembawang Shipyard Private Limited (2005), a subsidiary of SembCorp Marine Ltd., had collaborated with local schools by launching the "Green Wave" Environmental Care Project to develop young people's skills in investigating the environment and thus to foster them to acquire a concern and appreciation for the environment. This annual competition, which is open to students from across academic institutions, inculcates in students a sense of personal and group responsibility for their own environment.

Some companies have contributed to the public engagement in learning about the environment by funding young people to take environmental field trips abroad. For instance, Bayer, in cooperation with the United Nations Environment Program (UNEP), has organised and promoted a number of projects that aims to improve young peoples' knowledge about the environment and to encourage them in their commitment to doing something about the issue. One of these projects is the "Bayer Young Environmental Envoy Program", which selected students on a week-long study tour to experience how environmental protection is practised by people, government and industry in Germany about technology and environmental awareness (Bayer, 2010). Similarly, other companies like Hongkong and Shanghai Banking Corporation (HSBC) have started a Care-for-Nature program in 1991 and recently they have partnered with the National Youth Achievement Award Council to raise the local awareness of youth on environmental issues. They sponsored youths on a research study trip to the Puerto Rican rainforest on an Earthwatch expedition. The objective was to expand the students' knowledge about sustainable rainforest management. Puerto Rico was strategically chosen as it is identified to have successfully implemented a programme that has reversed its rate of deforestation through extensive reforestation (Cheam, 2010).

Most of the environmental programs undertaken by private organisations are innovative and are aimed to interest the public in environmental issues. Recently, an annual international campaign, Earth Hour, was held around the globe to raise awareness of climate change by asking participants to switch off their lights for a pre-determined hour. Local companies, including Star Hub, SingTel, Charles & Keith, Coca-Cola Singapore and Changi Airport Group (CAG), have also lent their support to help promote this annual event and the issue of climate change. Changi Airport Group (2011) set up exhibition panels, aimed to educate passengers and visitors about how they can protect the environment. Moreover, they also organised a simple quiz for passengers and visitors and gave an environmentally friendly, battery-less torch to those who participated in the quiz as a token to mark the occasion (Changi Airport Group, 2011). All these events and activities regarding environment protection organised by private sectors reported above have improved awareness and reinforced the idea of the Earth Hour.

Events

Public events are popular means for companies to make their advocacies known. In working with the Ministry of Environment and Sustainability, 63 partners organised over 130 activities during Climate Action Week in 2021 (Ministry of Sustainability and Environment, 2021). The National Trade Union Congress, through its youth wing and Singapore chapter of the global 350 organisation, campaigns for the reduction of current CO_2 levels to 350 ppm from the current level of 390 ppm. The campaign is set through a competitive run event called the NTUC Income RUN 350. As the title implies, the event supports the global 350 movement – to reduce carbon dioxide emissions to 350 parts per million by volume (National Trade Union Congress, 2013). The event that has been held since 2010 has attracted more than 10,000 runners by 2012. The event also donates part of the proceeds collected from registration fees and sponsorships to the Singapore Garden City Fund's Plant-A-Tree Programme. Their website reports that close to 630 trees have been planted over a three-year period from this initiative (National Trade Union Congress, 2013).

Recently an annual international campaign, Earth Hour, was held to raise awareness about climate change by asking participants to switch off their lights for a pre-determined hour. Local companies have also lent their support to help promote this annual event in Singapore. Such include the telecommunication companies Star Hub and Singtel, the apparel company Charles & Keith, as well as Coca-Cola and CAG. Further, Coca-Cola Singapore partnered with the Environment Challenge Organization Singapore for a campaign to increase awareness about climate change. Through this partnership's "Eco-ambassadors", 10,000 homes were reached in various housing estates. Some activities include sharing tips on how to adapt to changes brought about by climate change. Additionally, the CAG set up exhibition panels aimed at educating passengers and visitors to Singapore about Earth Hour. CAG also organised a simple quiz for passengers and visitors. Those who participated in the quiz received an environmentally friendly, battery-less torch as a token to mark the occasion. The student group Earthlink NTU of NTU also collaborated with the Orchard Road Business Association to be a part of Celebrate Earth Hour@Orchard 2011. All of these efforts were meant to raise awareness and support the Earth Hour event.

Elevating awareness is identified as the common objective of stakeholders championing climate change in their advocacies. While this addresses the crucial phase of establishing ground in driving home the message, it also raises the question of whether these initiatives facilitate deep and sustained understanding, or whether they encourage attitude and behaviour alterations. It is a topic in itself that warrants a separate investigation.

As a complement to classroom learning, informal venues open up opportunities for beefing up the climate change curriculum. It encourages cross-dimensional thinking about past and emerging issues encompassing the science as well as socio-political aspects of climate change. It also exposes the learner to real-life situations and to the different perspectives of stakeholders from the government, the civil and business sectors and local communities. Taken together, the combination of informal and formal modes promotes a deeper understanding of the phenomenon in a holistic and interdisciplinary way.

It is argued, however, that the informal curriculum is not a viable standalone alternative to formal education. By and large, the ad hoc nature of activities and networks lacks organisation and a coherent goal. The randomness at which these activities are organised results in a string of programs and projects that may be related to advocacy but are not interlinked so that there is widespread duplication and overlap. Learning in this way comes in a patchy, isolated, unsupported and unsustainable manner. Moreover, the selective nature of some programs may benefit only a few and deprives the rest of the cohort.

Collaborations and networks in the informal curriculum are loosely organised. One disadvantage of this feature is the absence of a body that checks on the accountability and accuracy of information taught. Education gained may fall short of quality to give way to corporate and organisational gimmickry and politicking. Indeed, the knowledge acquired may become limited to, and held hostage by, the agenda of the host institutions and organisations, thereby potentially misleading the learner. On the other hand, this non-rigid structure also presents prospects for innovation. The informal curriculum is a fertile ground to explore ideas and opportunities; the resources it is offering to add value to the formal curriculum are yet to be fully navigated and capitalised on. It is therefore paramount that a synergy is established between the two curricula, with the formal curriculum taking the lead and direction of learning.

How can public education bring about pro-climate behaviour?

While the examples cited above tend to be focused on raising awareness about the climate change issues, there is a desire to bring about learning of these issues beyond "learning to know" to "learning to do". In particular, we have established in the earlier chapters the desire for CCE to bring about changes to behaviour that are pro-environmental in nature.

Research has found that individuals were more inclined to pursue pro-environmental behaviours if their neighbours were taking climate-conscious measures. Wi and Chang (2019) looked at how CCE can encourage pro-environmental behaviour through Residential Committees (RCs) which are essentially grassroots organisations in the context of the study. The researchers were interested in how environmental education can improve individuals' understanding of the issue and motivate them to adopt pro-environmental behaviour?' This is done by providing people with the essential climate change knowledge, skills and values. Instead of just providing information in pamphlets, the team designed an educational programme called "Transformative Education for Climate Change" (TrEC) which included information packets, briefing sessions, question and answer sessions and household visits. This builds on the idea that students should "develop and undertake action based on a decision after recognising and understanding the issue" (Petersen & Barnes, 2020). A mixed-method study was performed on individuals in a precinct for three months with two check-in sessions to encourage and remind participants to practice conservational tips. The participants took part in a pre-programme survey and a post-programme survey after. A total of 173 participants were involved and the participants were split into two groups using purposive sampling, with 95 participants in the control group and 78 participants in the experimental group.

The program was delivered in three parts, utilising Mezirow's (1997) transformational learning theory: an introduction video, seminars and a question and answers session with the participants were provided in addition to an informational packet containing tips for conservation. Two visits to the participant's households were also made to check on their progress. The programme was administered for the experimental group with the control group only receiving the information packet. A test instrument to ascertain knowledge, attitudes and behavioural changes was administered for both groups before and after the TrEC programme. Further, the utility bills of the respondents were also collected to see if there has been a change in the energy consumption of each of the participants' households. In Singapore, the country in which the study was conducted, utility bills come with a summary of energy consumption patterns for the preceding months in the form of labelled bar graphs. This makes it possible to quantify the monetary impact of pro-environment behaviour due to the TrEC using the statistical tests deployed for this study.

Using Chang's (2014) conceptual framework to classify knowledge, skills and values as learning outcomes in connection to a person's understanding of climate change. The data obtained was contrasted and categorised into the following themes:

1. Knowledge
2. Attitudes
3. Actions
4. Monetary Savings

Parametric tests such as the Shapiro–Wilk normality test and the paired sample t-test were used to determine normal distributions and contrasts in utility bills before and after the programme. Surveys and discussion answers were coded into themes. The findings showed that the implementation of the TrEC programme proved to be effective in promoting pro-environmental action in individuals, at least in the amount of electricity usage. The experimental group had improved attitudes and a \$5.60 reduction in utility expenditures, whereas the control group had minimal change in attitudes and a $1.80 average rise in utility bills.

Conservational knowledge recommendations, on the other hand, revealed minute changes between the experimental and control groups. However, there was a considerable gap in certain knowledge items between the two groups. When individual values were analysed in relation to pro-environmental behaviour, it was discovered that the majority of participants demonstrated and self-reported behavioural change when it was linked to an individual's convenience and monetary benefit. There was also an observation that those who were better able to make decisions based on their environmental knowledge, we more motivated to adopt pro-environmental actions.

There were no significant differences in conservation knowledge between the two groups, indicating that control group members were aware of the steps to take to adopt climate-conscious activities. Participants who did not modify their behaviours stated a lack of awareness of the reasons or they did not feel the urgency to change their routines. As a result, the knowledge that focused on behavioural change – such as knowing why it is vital to do so – was critical. This explains why previous government initiatives were unproductive since they focused on raising awareness rather than motivating people to alter their behaviour.

Furthermore, when it was beneficial or convenient for them, individuals were more likely to participate in pro-environmental behaviour. Many participants backed this up by proposing incentives for such behaviour, describing how rewards (advantage) might inspire them to modify their behaviour. The study also found a newly emerging theme of "governance" within participants' motivations. Some participants felt that they should partake in these actions as it was knowledge and instruction given by the government and hence, trustworthy. Contrastingly, other participants did not feel obligated to participate in pro-environmental action as it was the State's responsibility to combat climate change. Essentially, education must be "transformative", going beyond awareness and providing avenues for individuals to establish and make sense of their experiences gained via activities, as well as relate to their beliefs.

Summary and discussion

As previously stated, the purpose of activities organised by informal sectors varies depending on the target audience (see Table 8.1). The activities designed to engage the general public with climate change issues mainly focus on transmitting the knowledge of climate change and communicating the importance of changing lifestyles to tackle the emergent issues brought about by climate change. A whole spectrum of learning outcomes that we have outlined in CCE is missing in this case. In contrast, the purposes of the activities targeting school and college students highly emphasise cultivating and enhancing young people's appropriate attitude towards the natural environment.

According to Table 8.1, it seems that there is a gap between the domain of knowledge and action in the activities targeting public groups, thus leaving people vague understanding of climate change and poor problem-solving skills to fight against its impacts on our future life. This is not dissimilar to the observations by other studies from aboard (Lowe, 2006). Similarly, in activities targeting school groups, the promotion of environmental attitudes is overwhelmed and leaves the mission of communicating knowledge, developing concepts and skills and changing behaviours behind. There is one exception, however – the national climate change competition introduced by NCCS which aimed at promoting young people's learning about climate change and growing awareness of the environment through identifying and studying climate change issues and the possible solutions. We acknowledge that public and independent organisations in Singapore have made great contributions to the public engagement in the battle against climate change. Nevertheless, it is necessary for the general public to understand the causing factors of climate change and how our life can be influenced by such an emergent issue. Therefore, the design of the CCE programs should highlight the conceptual understanding along with the attitude and action perspectives. Indeed, CCE for informal education settings should also be guided by the principle to *help learners develop knowledge, skills, values and action to engage and learn about the causes, impact and management of climate change.*

CCE has only begun to take root within the education system, partly because as an issue it has come to the forefront of public knowledge and debate only recently and partly because it is seen as a global problem that needed to be fleshed out into concrete issues that Singaporeans could appreciate as urgent and critical to its long-term future. The fact that most problems are handled and handled well by the government, means that the populace tends to be reticent and takes a back seat. "The government knows best" is an apt description of Singaporeans' perception of many issues facing the country. While the formal curriculum has in place topics that deal with the environment much of the awareness that school students obtain about environmental issues are obtained through the informal and co-curricular activities as well as through the various campaigns, projects and camps mounted by non-governmental agencies. It is only through a two-prong approach of both formal and informal curricula in CCE that Singapore will be able to ensure that the knowledge, skills and values so required by its people as applied to engage and respond effectively to impending changes in the natural climate system – changes that will occur at an unprecedented rate, scale and intensity.

Looking at the informal curriculum through an inventory of climate change-related initiatives by the government, civil society actors and the business sector, it indicates that the public including school students are predominantly exposed to awareness campaigns and a handful that target knowledge building, attitude and behaviour change. To advance the cause outlined here to imbue in public informal education the need to adopt a CCE framework that will guide the development of programmes to engage

Table 8.1 The purpose of activities organised by informal sectors for different target groups

Purposes	*General public*	*School/college students*
Knowledge	Climate change exhibition (Singapore Science Centre) Joint public forum (Singapore Science Centre, Singapore Institute of Biology) Program for Environmental Experiential Learning (Singapore Environment Institute) Educational forums (SEAGA) Eco-ambassadors (Coca-Cola Singapore) Environmental quiz (Changi Airport Group)	National climate change competition (NCCSS) National Youth Environment Conference (NYAA)
Concepts	Nil	National climate change competition (NCCSS)
Skills	Nil	National climate change competition (NCCSS) Green Wave Environmental Care Project (Sembawang Shipyard)
Attitudes	10% Energy Challenge (NEA) Care for Nature (HSBC)	Clean and Green Singapore Schools' Carnival (NEA) Youth for the Environment Day Program (NEA, MOE) National climate change competition (NCCSS) Environmental projects competition (TICE) National Youth Environment Conference (NYAA) NUS Fight Climate Change Carnival (SAVE) Earthlink NTU (NTU student group) Recycling station visit (Yishun Secondary School) National Weather Study Project Competition (SENOKO) Bayer Young Environmental Envoy Program (Bayer)

(*Continued*)

Table 8.1 (Continued)

Purposes	*General public*	*School/college students*
Action	TrEC programme (Wi & Chang, 2019) Everyday Superhero (NEA) 10% Energy Challenge (NEA) 1,000,000 Native Plants @ Southwest (CDCs) Change-A-Bulb @Southwest (CDCs) Switch Off @Southwest (CDCs) 25°C Campaign @Southwest (CDCs) Project E.A.R.T.H. (NEA)	National climate change competition (NCCSS) Recycling station visit (Yishun Secondary School)

various cognitive and affective domains of learning about the climate change issue, a special panel on CCE was organised in November 2012 at the Southeast Asian Geography Association Conference in Singapore. The ESD Programme Coordinator and Liaison Officer for the Southeast Asian Region at UNESCO Bangkok and the Director of the 3P (People, Public and Private) network at the NCCS, Singapore, were invited and presided as panellists at this meeting. While the conference was attended by geographers, social scientists and teachers from around the Southeast Asian region, a clearer understanding of the nature of the national and international initiatives for CCE was discussed. While this discussion forum was a kickstart to the professional dialogue between policymakers, NGOs, scholars and teachers, there is much to be done on the road ahead.

As a complement to classroom learning, informal venues open opportunities for beefing up the climate change curriculum. It encourages cross-dimensional thinking about past and emerging issues encompassing the science as well as socio-political aspects of climate change. It also exposes the learner to real-life situations and to different perspectives of stakeholders on the issue. Together with a well-designed CCE for school curricula, informal education on climate change can play a pivotal role in reinforcing the knowledge, skills and attitudes that will inform the action of our youth.

9 CCE, environmental education and sustainability education through geographical education

Much of the discussion in this book has focused on climate change education (CCE) while many systems, institutions and individuals consider CCE to be part of environmental education (EE) or education for sustainable development (ESD). We have explored the relationships among CCE, EE and ESD in Chapter 2 and explicated the overlaps and differences, but a gnawing issue of whether this schematic differentiation of discourses that seek to make the world a better place are artificially categorised and if such categorisation results in a less than holistic approach to ensure a better environmental future for us all. Regardless of how we define these concepts and how we design programmes to teach people about the issues organised around these concepts, we are all living on the same one Earth. Is there some way that we can integrate these approaches into existing curricular constructs?

Whether we are talking about rising sea levels in the discourse in CCE, EE or ESD, we are concerned with the same impacts on environment, economies and society brought about by sea level rise. Some may argue that to differentiate between whether the topic is taught in CCE, EE or ESD is being pedantic. To what extent will situating the discussion of the sea level rise topic within CCE, EE or ESD make a difference, say, in the price of fish? Theoretical purists will tell you that there are substantial differences, and one will be able to cite research studies that indicate how each epistemological approach will likely produce slightly different cognitive, affective and behavioural learning outcomes. A pragmatist will say that it depends on your classroom instruction and assessment and how well these have been designed to support students' learning, rather than whether you are teaching it for CCE or ESD. There is of course the option where the curriculum design is influenced by the conceptual framework within which it was developed – whether it is CCE, EE or ESD. Consequently, this has implications for how the teacher will develop instruction and assessment for the topic too. This chapter argues that there is already an existing curricular construct in the form of geographical education that could embrace the notions of CCE, EE and ESD while not diluting the intent and ontological essence of these approaches. However, we have to first identify a problem statement for this argument.

DOI: 10.4324/9781003093800-9

Teaching EE, ESD and CCE

Do we need to argue for a school subject that best integrates ESD, EE and CCE? While the substance to be learnt is the same, the problem perhaps lies in how the substance will be taught. And how well teachers can teach the material depends on how well prepared they are during their training. Teacher educators play a pivotal role in pre-service teacher programmes to help student teachers improve their own professional knowledge and practice in CCE and ESD (Corney & Reid, 2007). Through document analyses and interviews of the 22 student teachers and 15 mentors teaching geography in 15 schools in the United Kingdom, the subject matter knowledge, pedagogy and the sources of their understanding of sustainable development were examined. The study showed that both the subject matter knowledge and professional pedagogical knowledge contribute to the student teachers' professional learning and support ESD. Student teachers demonstrated their mastery of sustainable development by clearly articulating its complex and contentious notions. They were also able to articulate the extent to which they understood the relationship between the subject matter and ESD pedagogy, which included spatial inquiry and constructivist pedagogies. Students credit their ESD learning to collaborative efforts between schools and teacher trainers in including diverse activities in the classroom context and within their pre-service training. For example, between the mentor and the trainee teacher, there is a first conversation about different methods for ESD. This is followed by lesson preparation for student teachers and a debriefing of student teacher teaching with the mentor and the university tutor (Corney & Reid, 2007).

A separate difficulty develops in a piece of research conducted in the United States, when the student teacher's subject matter expertise is missing in the conceptualisation of a planned class. The existing education system, driven by "performativity and accountability", places serious restrictions on the potential of ESD (Firth & Winter, 2007). There was little indication that concepts of sustainable development were employed in the targeted learning objectives in semi-structured interviews with four student teachers. Instead, they focused on students' personal responses to environmental and development challenges in their daily lives, taking into account what they can do as people. While the student instructors possessed extensive topic expertise, they were more concerned with performing and delivering the predetermined curriculum than with capitalising on their own subject knowledge. In this scenario, an atmosphere of performativity and accountability in teacher education was linked to the student teachers' tenacity in sticking to the intended goal and their reluctance to apply their subject matter expertise to adapt and engage their pupils (Firth & Winter, 2007).

Given recent developments in UK education policy, Summers, Childs, and Corney (2005) examined transdisciplinary approaches to ESD in initial teacher training (ITT) partnerships. They discovered that student teachers have a greater comprehension of the notion of sustainable development than their mentors using questionnaire surveys of ITT partnering schools, geography and science student teachers, and their school mentors, and that the "geography mentors perceive themselves to be better prepared for mentoring in this area than their science colleagues" (Summers, Childs, & Corney, 2005). Indeed, mentors believed that a lack of "subject expertise" was the primary reason for them feeling unprepared to help student instructors. To that end, the study underlines the need for professional development aimed at strengthening mentors' grasp of the issue and necessary pedagogical skills in order to better assist ESD implementation.

According to an Organisation for Economic Co-operation and Development (OECD) assessment of policy development in environmental education for Australia, Austria, Finland, Germany and Norway, pre-service education has been identified as the Achilles' heel in environmental education (McKeown-Ice, 2000). There has been no systematic national or regional assessment of environmental education as a component of pre-service education in the United States prior to 2000. McKeown-Ice (2000) found that faculty motivation and expertise, as well as state certification laws, were the two causes that resulted in the situation in pre-service for Environmental Education in a survey of 446 institutions. Furthermore, there was limited course time, as the lessons were usually integrated into coursework, primarily into science methods courses. Although a few tertiary institutions in the United States provided an environmental education course, focus or even a stand-alone module, few emphasised the aims of environmental education via action. In terms of faculty capability, around a third of the responding schools had an environmental education specialist on staff, while the other half had none (Mckeown & Hopkins, 2010).

In order to examine how Environmental Education is best incorporated within the curriculum of teacher preparation programmes Heimlich, Braus, Olivolo, Mckeown-Ice, and Barringer-Smith (2004) surveyed 449 institutions across the United States. The study wanted to uncover how Environmental Education is taught within teacher preparation programmes, the scope of issues for Environmental Education curricula, perceived resources for and barriers to Environmental Education and the relationships between perceived barriers and Environmental Education course concepts. More than half the institutions offer a course on Environmental Education and only 14.8% of the respondents offer Environmental Education by incorporating it into an existing teaching methods course. Further, the extent that various environmental concepts are included in teacher preparation programmes is very low (mean scores ranged between 2.4 and 2.8 out of the 5-point scale), which indicates that these topics are not offered in required courses for the majority of the respondents (Heimlich, Braus, Olivolo, Mckeown-Ice, & Barringer-Smith, 2004). In terms of barriers, time and the perceived lack of resources were highlighted. Evidently, the National Council for Accreditation of Teacher Education (NCATE) requirements are an important consideration for the incorporation of Environmental Education into teacher preparation programs. Also, trans-disciplinary Environmental Education is process-skill in nature, rather than disciplinary-based. Incorporating environmental education into existing teacher-preparation programs is perhaps most impeded by political agenda; in turn, this poses a challenge to institutionalise Environmental Education as it requires collective action. The key impediments of certification and accreditation may indeed offer a potential political tool for Environmental Education to be incorporated into the accreditation and national certification efforts (Heimlich, Braus, Olivolo, Mckeown-Ice, and Barringer-Smith, 2004).

Across the Atlantic, in a study of two teaching training colleges in Belgium, Van Petegem, Blieck, Imbrecht, and Van Hout (2005) investigated the implementation of environmental education. Nearly all Flemish secondary schools provide some Environmental Education but, in most cases, a synergistic and holistic approach is lacking. Indeed, Environmental Education is absent in teacher education. While colleges and universities do provide Environmental Education, it is commonly limited to topics on ecology in geography or biology. Through interviews, questionnaires and focus group discussions, the key constraints for implementing Environmental Education in teacher education programmes include the lack of knowledge, a reluctance to expand into new fields of study, a reluctance of non-science teachers, to teach Environmental Education as feel that it was the science teachers' responsibility, as well as a lack of open-mindedness for innovative didactic competencies. The study suggested that there was a need to showcase success stories to inspire teachers and to encourage educators to participate in the policy-related activities for Environmental Education (Van Petegem, Blieck, Imbrecht, & Van Hout, 2005).

In order to better understand how Environmental Education can be infused into the educational system, Powers (2004) studied the perspectives of faculty who educate pre-service teachers. Through telephone interviews with 18 faculty members who teach "at least one social studies or science methods course each year and is known to incorporate some aspect of Environmental Education theory and practice into his or her teaching" (Powers, 2004), the study investigated how Environmental Education theory and practice are incorporated into the methods course, uncover the barrier to implementation and possible mitigation of these impediments. The interviewed faculty prefers the infusion of Environmental Education into existing methods courses over offering it as a separate course, although limited curriculum time was considered the greatest limitation to infusing Environmental Education, in addition to students' repulsion from science. The faculty also perceived that it is important to initiate the integration of environmental education at the pre-service level, in that it encourages teachers to reflect on how to teach social studies within the context of the local environment (Powers, 2004).

Student teachers often have their own views about the content and how to teach it, which would consequently influence "their thinking about teaching and their classroom practice" (Corney, 2000, p. 313). At the beginning of their teacher education programme, student teachers have their own ideas about what teaching and learning entails, but these pre-conceptions are influenced by a range of prior experiences, which do not appear to change much during the training programme. Through the collective case study of three geography student teachers, the study by Corney (2000) describes geography student teachers' thinking of and practice in teaching environmental topics and uncovered the influences on their thinking and practice. Firstly, the student teachers have detailed and well-developed geographical content teaching about the environment. However, the pedagogical pre-conceptions were "less detailed and less clearly justified" than those of subject matter (Corney & Reid, 2007). An implication is that pre-service programmes should encourage student teachers to reflect on the subject-specific pedagogy. The influence of a student teachers' prior education from secondary school to university as well as from informal education settings is evident, and prominent within these are the influence that particular teachers have on them and the nature of curriculum and type of courses that student teachers taken (Corney, 2000).

In the previous studies reviewed, the subject of study was all pre-service teacher education. The author has only found one national survey that examined the state of Environmental Education in-service education in the USA. The results showed that professional development in Environmental Education is commonly conducted as activity-based lessons and based on nationally produced curricula (Wade, 1996). The in-service courses are designed based on the science subject rather than interdisciplinary in nature. The courses are concerned more with environmental subject matter than educational context. This raises the need to have effective communication of the interdisciplinary nature of Environmental Education to all stakeholders of the formal education community and to begin doing so by examining definitions of Environmental Education (Wade, 1996).

In summary, the literature reviewed on teacher education for ESD or EE tends to be concerned about whether the teachers have sufficient subject matter knowledge, if the context for EE or ESD was taught as an independent subject or integrated with a methods course and if there was sufficient avenue for the teachers to upgrade and update their subject matter and pedagogical knowledge. This is not unlike the findings for Singapore (see Chapter 6) in which teacher focus group discussions uncovered a strong demand for teacher professional development platforms to engage on the issue of climate change.

Given the overwhelming concern for subject matter knowledge and where the discussion of CCE is situated in the school curriculum, I argue that the disciplinary domain that best matches the type of training required will be Geography. Consequently, the proposition in this chapter for the school geography subject as a natural avenue to teach about CCE, EE and ESD becomes a viable solution.

The purpose of geographical education

The International Geographical Union Commission on Geographical Education (IGU-CGE) developed the International Charter on Geographical Education in 1992 and it was endorsed by the International Geographical Union General Assembly at the 27th International Geographical Congress in Washington, D.C., USA (International Geographical Union – Commission on Geographical Education, 1992). Its purpose was to inform policymakers of the significance of "teaching and learning Geography for the well-being of a country and its people, as well as that country's capacity to function in a global age" (Stoltman, 1997). It also sought to instruct curriculum planners about the objectives of geographical education and "the general standards to which it should aspire". The 1992 Charter was updated in August 2016 to reflect developments over the years and the direction of geographical education. The 2016 Charter emphasises worldwide collaboration, idea exchange and curriculum comparison, and it was presented "as an action plan" (International Geographic Union – Commission on Geographical Education, 2016, p. 2) and has been a guide for many geographic educators in terms of research, curriculum making and teaching. The 2016 Charter suggests that "geographical education is indispensable to the development of responsible and active citizens in the present and future world" (International Geographic Union – Commission on Geographical Education, 2016, p. 1).

In fact, the charter proclaims to support the principles set out in the UN Sustainable Development Goals (International Geographic Union – Commission on Geographical Education, 2016, p. 1) and affirms the Lucerne Declaration on Geography Education for Sustainable Development (International Geographic Union – Commission on Geographical Education, 2016, p. 2). The Lucerne Declaration (2007) was made at a time shortly after the UN Decade of Education for Sustainable Development 2005–2014 was introduced. The assertion is that Geography is "a vital subject and resource for 21st century citizens living in a tightly interconnected world. It enables us to face questions of what it means to live sustainably in this world". The subject also allows people to learn to live together harmoniously with each other and with other living species, by helping people "to formulate questions, develop their intellectual skills and respond to issues affecting their lives" (International Geographic Union – Commission on Geographical Education, 2016, p. 5). Geographical education also helps learners deepen their understanding of many contemporary challenges such as food security, overexploitation of natural resources, climate change, overpopulation and urbanisation, which are all relevant to issues to include in ESD.

To further illustrate this alignment, the goals of ESD are consistent with the concept of universal education in developing children to be critical thinkers and informed decision-makers (Dewey, 1997; Jucker, 2011; Caiman & Lundegård, 2014). Geography education allows teachers to design learning based on disciplinary knowledge, as well as to underpin the interdisciplinarity of knowledge and ESD pedagogies through geography (Eilam & Trop, 2010) to achieve its expected outcomes (UNESCO, 2012). Chambers (1999) suggested that an important area in environmental education is "… learning how knowledge, values and attitudes can be transformed into action for the environment whilst at the same time avoiding indoctrination" (p. 92). I argue that such is the case for ESD in Geographical Education as well. In teaching ESD, geographical education approaches such as place-based education (Casey, 1993), bio-regional nature immersion (Hensley, 2011), eco-literacy (Goleman, Bennet, & Barlow, 2012) and powerful knowledge (Young, 2008; Roberts, 2014) are fundamental. Haubrich (2000) went on to claim that the discipline of geography provides the knowledge and skills necessary to grasp the growing realities of a globalised society in which resolving current concerns of sustainable development is critical. Indeed, geographical education can "contribute to the joint goals of personal happiness and the sustainable development of the world" (Haubrich, 2000, p. 284).

Geographical competencies for EE, ESD and CCE

While the argument thus far has focused on international charters and declarations, reflecting the aspirations and discourses of international professional organisations of the school geography curriculum, the academic argument for geographical education to bring EE, ESD and CCE together must be developed further. By examining the Lucerne Declaration, one can surmise that it argues that the key geographical competencies required to support ESD are in geographical knowledge and understanding, geographical skills and attitudes and values. Incidentally, Environmental Education discourses have also focused on the knowledge, skills and attitudes triad. In Chapters 2 and 3, the learning outcomes for CCE and the proposed framework of knowing, doing and being are also in alignment with the Lucerne Declaration argument.

Indeed, the geographical knowledge and understanding required in ESD comprise knowledge of the Earth's primary natural systems in understanding the interactions within and across ecosystems. It also includes the understanding of the socio-economic systems. Geography's unique treatment of spatial concepts allows our learners to make sense of the world through "location, distribution distance, movement, region, scale, spatial association, spatial interaction and change over time" (Haubrich, Reinfried, & Schleicher, 2007, p. 245). Knowledge of our human and natural systems over space is critical in understanding the issues around sustainable development and finding ways to solve the related problems. For the climate change topic, in particular, this geographical knowledge and understanding are required before a learner can make sense of the issues in the causes, impact and management of climate change. Geographical education helps students know and understand climate change.

The Lucerne Declaration also discussed the geographical skills used in communicating and thinking about topics that range from the local to the international scale. Implementing global ESD goals into state policy has clearly become more common in the last decade. According to Olsson, Gericke, and Chang-Rundgren (2016), SD must be explained for each of its dimensions in terms of their interrelationship in both time (past–present–future) and space (near-far). There is also the inclusion of hands-on and social skills needed for learning about sustainable development issues. In the context of CCE, asking questions of what, where, when, how, who and why in relation to geographical thinking can be supported by field inquiry, for example. These feature strongly in CCE as well as geographical education, in terms of the pedagogies and assessment discussed in the previous chapters.

ESD encourages children to be active members of their communities. It is a transformative approach that develops students as problem solvers and action takers in their own context. A student who may begin with "embryonic geographical knowledge and understandings" (Catling, 2005) and "think of things as connected, even if he/[she] revises his[/her] view with each succeeding year, has begun the life of learning" (Van Doren, 1959 p. 155). Geography education endeavours to help students in developing their "dedication to seeking solutions to local, regional, national and international questions and problems" (Haubrich, Reinfried, & Schleicher, 2007, p. 245) – being educated about climate change.

In summary, the Lucerne Declaration argues that "Geography Education can greatly contribute to achieving the goals of the United Nations Decade of Education for Sustainable Development by providing relevant knowledge, skills, values and attitudes crucial for a peaceful coexistence of individuals with nature on this planet" (Haubrich, Reinfried, & Schleicher, 2007, p. 246).

CCE and the school geography curriculum

While this chapter sets out to discuss how CCE, EE and ESD can be carried out through the school geography curriculum, the focus of the book is still on CCE. The question that remains is how CCE can be integrated into the school geography curriculum, the instruction and its assessment. Teachers' involvement as curriculum makers in addressing the context of learning is critical to the development of learners' geographical thinking about climate change.

Teachers, according to Lambert and Hopkin (2014), are key stakeholders in the curriculum-making process. Even if a state-mandated curriculum exists, instructors make daily judgements about topic sequencing, materials selection, learning activity design and assessment. The notion is that for geography to be taught and learnt effectively, the learners' experiences, the teachers' decisions and the body of knowledge that makes up classroom geography must be independent or mutually exclusive. In other words, how well a teacher knows the learner's profile, the teacher's subject matter expertise and the instructional methods to be employed determines how effectively the student will learn (Lambert & Hopkin, 2014).

In the case of CCE, we must take into account the student's "perceived needs and our understanding of learning". Teachers then select "the best of what we know and inducting students into the processes and procedures of how we have come to know it" (Lambert & Hopkin, 2014, p. 65). The teacher plays an important part in the curriculum development process. Indeed, the intersection of the three domains of geography curriculum making in Figure 9.1 necessitates the instructor carefully considering which learning activity to use, what the essential ideas are and how they might assist students to think spatially to go beyond what they currently know.

Geography is concerned with asking questions of:

1. Where is it?
2. What is it like?
3. Why is it there?
4. How did it happen?
5. What impacts does it have?
6. How should it be managed for the mutual benefit of humanity and the natural environment?

(International Geographic Union – Commission on Geographical Education, 1992)

Using these questions, we can easily frame and include concepts in CCE, EE and ESD but it does not magically get transformed into classroom practice. In fact, things like how the lessons are sequenced, what activities to design,

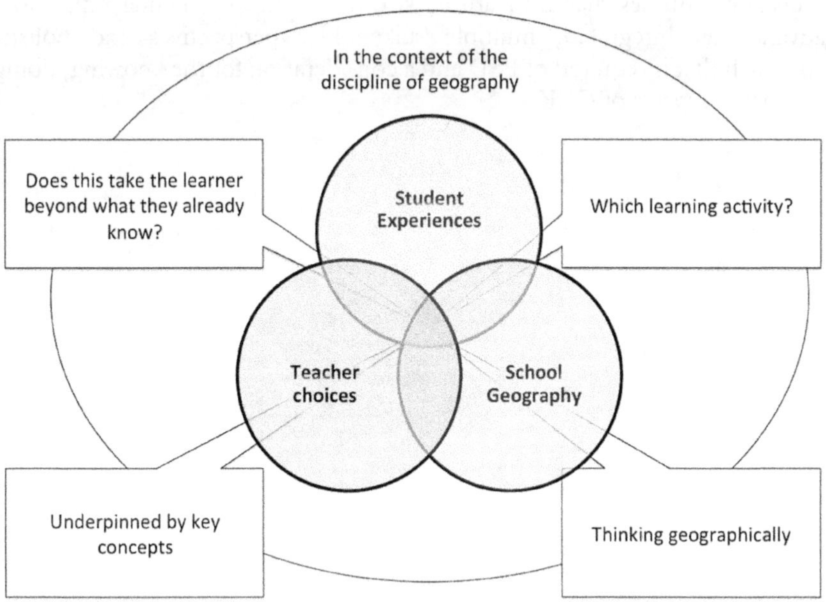

Figure 9.1 Curriculum making in Geography (Lambert & Hopkin, 2014).

which resources to choose, what technology to support the learning and how they would assess learning, might be extremely intimidating for a new instructor. A teacher acquires profound pedagogical content knowledge (PCK) via experience and ongoing professional development, allowing him or her to construct a meaningful learning activity (Shulman, 1986). Teachers use their subject discipline expertise to analyse the curriculum, discover resources that best reflect the concepts in the subject matter and construct learning activities that assist students to bridge the comprehension gap. PCK is required for the curriculum, teaching and assessment cycle since learning cannot occur outside of this environment. Furthermore, topics in CCE require an

> awareness of common misconceptions and ways of looking at them, the importance of forging connections among different content-based ideas, students' prior knowledge, alternative teaching strategies, and the flexibility that comes from exploring alternative ways of looking at the same idea or problem are all essential for effective teaching.
>
> (Koehler & Mishra, 2009, p. 64)

In summary, the school geography subject allows a teacher to integrate concepts of EE, ESD and CCE easily through the curriculum-making process. In addition to the alignment in intentions as outlined in the Lucerne Declaration, the subject matter such as topics, issues, solutions, challenges and debates can be taught through different subjects, and it is with the geography subject that teachers can take tackle the dimensions of knowledge, skills and attitudes that are featured strongly in Environmental Education, advance an integrated, multiple stakeholder perspectives and holistic approach that is required of ESD and a consideration for the knowing, doing and being aspects of CCE.

10 Is CCE relevant for the global citizen in these critical times?

Climate change on Earth is an unprecedented phenomenon in terms of rate, scale and complexity. Even as you read this book, new reports will be released by intergovernmental panels and people debate over coffee about whether climate change will impact our lives negatively. As an educator living in extraordinary times, one cannot ignore the social responsibility to advance educational discourse on the matter. While climatologists like James Hansen have devoted their academic excellence to advance our knowledge about the climate change phenomenon, climate educators in their role as applied climatologists, have a similar responsibility to translate the research findings for non-scientists in vernacular terms so that we can all play a part in our common environmental future. This final chapter will summarise the key ideas presented with special emphasis on the importance of climate change education (CCE) for both the present and future generations.

In discussing the topic of this book, four key questions come to mind. Firstly, what is the purpose of CCE? Secondly, what do we teach in CCE? Thirdly, how do we teach the content and subject domain of climate change? Finally, how do we know if the student or a person has learnt something about climate change? To this end, each of these purposes would resonate with different audiences. In answering the purpose of CCE, policymakers, educators, educational researchers, teachers, students and even the person on the street will be interested. This of course is informed by the various climate change discourses which have been discussed in the book. The section in Chapter 2 on what constitutes CCE reconceptualises the way climate change content can be organised and thus the design of the curriculum. This offers a framework that was not previously available for defining what people mean when they state that someone has learnt about climate change. In terms of the instructional methodology to answer the "how do you teach" question, it is probably the most relevant section to practitioners in the field, such as teachers and even educators in the informal education setting. Of course, the section in Chatper 7 on assessment of "how can we tell if someone has learnt about climate change" will also be relevant to practitioners.

DOI: 10.4324/9781003093800-10

Uncertainty and debates about climate change

The classroom is purportedly the best venue to start a sustainable climate change mitigation strategy, both in the short- and long term. However, CCE has often been derided as an instrument to advance the discourse of climate alarmists (The Heartland Institute, 2007). This is unsurprising given the layperson's perception of fiery debates over whether climate change is indeed taking place or if human-induced climate change will have a definite impact on society and economy, given the complexities and uncertainties embedded deep in the climate change discourses. Unfortunately, an informed understanding of the phenomenon is hindered by the complexity of the science of climate change. This is made worse by the complexity of reading news about global changes while having to make sense of local changes to the environment. Further, the public is unable to discern the observable impact of climate change over the short term. Human memories of weather events tend to be limited to recent occurrences unless the event is exceptionally catastrophic. So perhaps Ungar's (1992) claims that the environmental issue has to be associated with dramatic real-world events in order for it to gain notice is true. However, I have noticed first-hand how residents of Singapore were quick to work themselves into a frenzy over the occurrence of hailstones in June 2013 when the last occurrence was just less than five years ago (National Environment Agency, 2013). While the local bureau called it a rare occurrence, people said they have never seen any hail in Singapore, nor recalled it ever being reported on the news. The vexing aspect of this is that hailstones are indeed a dramatic event, yet people seem to only remember the most recent one. The issue of climate change also loses its novelty "celebrity status" as media coverage wanes. Subsequently, the interest levels are worn down to ambivalence due to the long lags between observable weather phenomena.

As we examine the various discourses of climate change, what becomes clear is that while the scientific, political and economic communities have much to resolve, the ethical concerns for the environment can be addressed independently of the supposed debates. Uncertainty, complexity, vulnerability and, ultimately, the authority and validity of science-based policymaking are predominant themes in the scientific discourse of climate change. The economic discourse of climate change is built on the concepts of costs and benefits of mitigation versus adaptation. The concept of uncertainty affects the scientific discourse on the causes of climate change and the economic discourse on the costs of climate change, and the political discourse results in varying degrees of adoption of adaptation and mitigation strategies. Finally, there is the ethical discourse of climate change where responsibilities and accountability form the foci of climate change issues. Learning about climate change need not mean that a student should subscribe to only one side of the alarmist–contrarian divide. Indeed, the fundamental premise that drives learning about climate change should be that our children are custodians of our common environmental future, and that they have a mandate to critically engage in the issue and act based on well-informed decisions. Consequently, I argue that CCE should be built on the foundations of moral responsibility to the environment.

Framing our understanding of climate change education

While there is a general consensus that people need to learn about their environment, the field of CCE is relatively new and using the framework for EE and ESD will not immediately engage the divergent discourses in climate change. A robust framework which recognises the range of subject domains as well as varied cognitive and affective engagement is needed. Of course, the book has also argued for the school geography subject to be the avenue through which CCE can be implemented. To this end, the question of what does one mean when we say someone has learnt about climate change was posed to derive a conceptual construct that can answer this. Naturally, the works in educational psychology on educational objectives and learning conditions were referred to in the analysis. The conceptual construct also took into consideration the domains of climate change issues from the causes to the impact and the management aspects. When all of these different aspects of what it means to learn about climate change were brought together, the result was the development of a framework that covers the three dimensions of climate change discourse, domain knowledge, and cognitive and affective engagement. Indeed, *climate change education seeks to help learners develop knowledge, skills, values and action to engage and learn about the causes, impact and management of climate change* – developing a learner's knowing, doing and being in relation to the climate change issue.

The tenets and principles for teaching climate change in the resultant framework do not defer from those of ESD and EE. Indeed, the proposed framework addresses a range of engagement, from awareness to knowledge and ultimately to participation, and can be found within the school geography subject. Further, issues of environment, society, culture and economy are locally relevant and culturally appropriate, address content and build civil capacity for environmental stewardship and good quality of life. Indeed, the framework complements the work of ESD and EE while targeted specifically at the teaching and learning of climate change. The framework describes that CCE can help students learn about the causes, impact and management issues. It also clarifies if students are expected to recall or understand the knowledge, use skills such as analysis and evaluation, take a stand or even take action to do something about the state of the environment.

To operationalise this framework, an analysis of the Singapore school curricula was undertaken. While the key aim of which was to understand where CCE exist within the formal curriculum, it also afforded the opportunity to articulate through the example of the Singapore curricula, the potential applicability of the framework to other curricula. An argument has also been made for the school geography subject to be one key avenue to bring together the various approaches of EE, ESD and CCE as the subject matter to be learnt does not differ but Geography as a subject discipline allows us to teach these various approaches holistically as we carry out the school geography curriculum.

Tacitly the framework assumes that not all students are at the same starting point in learning about climate change. Indeed, the students are all at varying stages of readiness and in order for the teaching and learning to be effective and meaningful, it is pertinent to understand what students know.

Determining student readiness in learning about climate change

An extensive literature survey for studies around the world and the case study for Singapore were referred to, in examining the baseline understanding of climate change issues. Although Singapore students claim to have received a great deal of information on climate change, there are gaps in parts of their understanding of climate change. Indeed, students even have misconceptions about the topic such as the belief that the ozone hole contributes to global warming. Other misconceptions include the confusion between the role of UV rays and infrared rays in greenhouse warming and the belief that natural disasters such as tsunamis and earthquakes are caused by climate change. The discouraging fact is that these findings locally are congruent with the studies conducted worldwide.

Students in general say that they are worried about climate change, but when it comes to taking action, they tend to be hesitant. In fact, they remain ambivalent in encouraging their friends to adopt pro-environmental practices. While a good number claim they are consciously reducing energy usage like turning off the lights when leaving a room, the interviews reveal that such behaviour could best be attributed to habits or parental control.

In all, teenagers in Singapore, at least those in lower secondary levels, view climate change from a neutral, distanced perspective – a discovery that is not dissimilar to the findings of Lowe (2006). They do not position themselves on either of the extremes (activism–apathy) but are bent towards having a neutral, ambivalent stand. Indeed, there is a negative correlation between their own perceived knowledge to both the attitude of involvement and engagement in learning. Students who are disinterested to learn about climate change are more likely to believe that their actions will not have any effect on the planet. This is similar to Trumper's (2010) observation that students in developed countries tend to lack concern about environmental issues such as climate change.

Perhaps formal curricula will help learners bridge the gap between knowledge, understanding, beliefs and action. The framework for CCE indicates a progression pathway for learners to move beyond knowledge recall to evaluation and even action. A chapter (Chapter 5) on pedagogies that accompany what students need to learn has also been included to support teaching and learning in CCE. In order to implement the CCE framework well, teacher preparation is crucial, as CCE does not exist as a separate subject in most education systems. In fact, a literature review shows that most teacher education programmes teach climate change as a component of the methods course.

Preparing teachers to teach climate change

Some teachers consider that climate change is a very difficult topic to teach, especially when the discourse is shrouded in debate and when there exist many misconceptions among students. While students have difficulty comprehending the difficult and complex concepts, some teachers have admitted that they struggle with the content as well. Findings from a survey and a focus group discussion on how well teachers think they are ready to teach climate change were included in this book to provide the reader some real concerns in the classroom. While subject matter knowledge was an area that teachers felt that students lack, they also commented on the lack of teaching resources or professional development activity. Teachers also commented on how students tend to remain ambivalent on the subject and that the support from the students' families is missing in reinforcing and sustaining the students' positive actions for the environment. The issue of assessment was also raised and there was a general call for more climate change to be included in formal curricula.

The literature suggested that one way that teachers could be prepared to teach climate change was to help them convey accurate information and instructional strategies that will help students hone their skills in thinking (Mckeown & Hopkins, 2010). Indeed, teachers should help students understand the complexity of climate science so that the child can adopt a critical approach to analysing and evaluating the mitigation and adaptation strategies with a view to engaging the socio-scientific debates (Schreiner, Henriksen, & Hansen, 2005). A key pedagogical approach to using refutation instruction was introduced through the work of Chang and Pascua (2016) and Chang, Pascua, and Ess (2018) to describe how climate change misconceptions can be corrected. Apart from helping students develop accurate knowledge and deep understanding, teachers should encourage responsible attitudes and be inclusive in engaging scientific, societal, political, ethical and personal discourses. Indeed, the teacher should empower the learner to realise that he/she is a stakeholder in the environmental future of Earth (Schreiner, Henriksen, & Hansen, 2005).

Teachers normally emphasise mitigating climate change through changing student behaviour when they think about how to teach it. Moreover, few teachers are able to articulate the best ways to teach about the subject matter and suggest pedagogical approaches. Certainly, a teacher must have a good understanding of *what* and *how* a topic should be taught before he or she goes into the classroom. To this end, teacher professional development workshops were discussed in this book. The first was concerned with providing teachers an avenue to examine their own subject matter knowledge through a conceptualisation exercise. The intentional focus on subject matter knowledge was to help teachers determine what they need to teach. This was an important part of reading the curriculum, interpreting the curriculum and then deciding what to teach for the curriculum. The framework advanced in this book was introduced at the workshop and teachers had to build their own frameworks through concept mapping as a means to consolidate their own subject matter understanding of the topic. In some sense, it helped the teachers determine what they knew and what they did not. Subsequently, a second workshop conducted subsequently focused on how the teachers could teach subject matter knowledge through a Conceptual Change method. Through the workshop, teachers were able to critique existing lesson plans and improve them to teach for conceptual change and help students clarify misconceptions. In essence, the teachers were able to develop an approach to critique and improve the instruction for CCE. Although well-prepared teachers are necessary for a successful implementation of CCE, the framework only articulates the key learning outcome categories. How can the curriculum and instruction be tied with the assessment of a climate change?

By using a backward design curriculum development approach (Wiggins & McTighe, 2005), assessment for learning is first considered before thinking about the curriculum design. This pragmatic approach requires first a clear articulation of what we would want our students to learn about CCE. This is informed by the conceptual framework that is developed in this book. Subsequently, the evidence of learning is determined by taking reference to the specific parts of the framework. Next, a performance task is designed that will allow the student to show that they can attain the evidence you are looking for. The task design will inadvertently require the teacher to consider the instructional approaches. Central to this approach is that it ties the interpretation of a national-level curriculum to the classroom through the assessment task. By using a method of performance assessment which is based on authentic tasks such as activities that require students to show what they are capable of, it ties back to the learning outcomes for the curriculum and the instruction for CCE. Indeed, for CCE to be successfully implemented, all three components of curriculum, instruction and assessment must be considered within the framework of learning outcomes described in this book.

Unfortunately, there still exists a frustrating issue of the reported gap between student's attitudes and actions. To make matters worse, students tend to have scientifically unusual but inaccurate explanations about climate change. Mass media and public discourse have been blamed as key sources of informal learning among students. In addition, with the limited curriculum time set aside for climate change, students find a ready source of learning from the mass media and even social media in some cases. This phenomenon is worrying as the coverage on the news and social media tend to be sensationalised and students seldom validate claims made on personal blog sites, for example. Indeed, how informal education can educate our children on climate change cannot be overlooked.

Public informal education to supplement formal education

The chapter on informal education (Chapter 8) has shown that there is also a gap between the domain of knowledge and action in awareness campaigns organised by public groups. It often creates a nebulous misunderstanding of the climate change issue and does not elucidate the reasons for the actions taken, often leading to un-sustained practice. In the same vein, public education aimed at school groups tends to omit knowledge and the understanding of concepts and focus on developing awareness. While informal education is highly dependent on the context of the society in which it is found, studies from around the world have uncovered the lack of accurate knowledge as the key factor to un-sustained environmental behaviour change. However, the Singapore example of the TrEC programme (Wi & Chang, 2019) provided a glimpse of how public education can be intentionally designed to encourage pro-environmental behaviour of the public.

Nonetheless, it is crucial for the general public to understand the causes and impact of climate change and CCE programs should highlight the conceptual understanding along with the attitude and action perspectives. To some extent, I am calling for the adoption of the CCE framework of learning outcomes developed in this book as a blueprint against which national and international agencies can coordinate public education efforts within the country and possibly in the region.

While CCE is in its infancy when compared to her EE and ESD cousins within the school systems in many countries, the formal curriculum has the potential to leverage the amount of information students learn about environmental issues through informal and co-curricular activities. A good example of how collaboration between agencies has resulted in meaningful development of CCE in Singapore is the intensive dialogues between the NCCS and the Ministry of Education of Singapore. The collaboration resulted in the reintroduction of an entire topic on varying weather and changing climate to the Geography subject for Secondary 3 and 10. Further, the topics covered range from knowledge to skills, values and action, set against the content of the causes, impact and management of climate change issues. It is only through a holistic approach of both formal and informal curricula in CCE will ensure that people have the knowledge, skills and values to respond to impending changes in the natural climate system.

Résumé – action for our common environmental future

In writing the concluding remarks for this chapter, I looked up the Wikipedia™ entry on education. Just like our students who turn to mass media and the web for information because it is convenient and ubiquitous, I chose a quick click and search over walking to my bookshelf. The word "education" is derived from the Latin *ēducātiō*, meaning to breed, bring up or rear (Wikipedia, 2013). As I reflected on using the analogy of a parent's advice to a child on food wastage in discussing the ethical discourse on climate change in Chapter 1, two things come to mind. The first is that we as educators have to walk the talk. If we encourage our students to verify facts beyond what they can find on the Internet, then so should we. Similarly, if we impress on students the need to learn beyond facts and understanding to doing something about our environment, then so should we. Learning about climate change does not stop at finding out about something. I will argue that it does not even stop at doing something about it. In one of the concept maps drawn by the teachers, right at the centre of the diagram is the term "being". CCE is about living our lives as climate enlightened beings. This might sound trite to an academic reader but in essence, what educators aspire, is to see an individual live his or her life meaningfully. When people ask me what I teach, I will not be saying I teach climate change. In truth, I teach people.

The second is that education always assumes that there is a future. To provide an example, we teach the next generation, and many people see education as a strategy for ensuring a sustainable future. The sad irony is that while the environment is not bothered about climate change, neither are we. Let me clarify. Climate change has been occurring throughout the Earth's natural history and it is continuously happening. Even if global temperatures continue to climb, the climate system will reach a new state of dynamic equilibrium. To Earth, this is merely another swing in the periodic ups and downs of world average temperatures. Certainly, the ecosystem will be affected and mass extinction may result. The irony is that humans should be really worried, but not everyone is concerned enough. Because even if this change was not brought about by human activity, it will affect our future. If our students are ambivalent about taking action, there will be little prospect of a liveable future for them. More importantly, students must learn to be critical thinkers at the interface between scientific knowledge, environmental activism and political discourse. It is not sufficient to just act blindly about our environmental future but there is a need for our children to engage in the issues, debate and then act on them.

The normal way to approach CCE is by considering causality as the organising concept for **why** we need to study climate change while morality is used as a meta concept to organise **how** we teach people about climate change. However, I posit that we should start from the concept of morality in addressing the **why** or the purpose of CCE and use causality as the organising concept for **how** to change our actions and behaviour to mitigate the impact of climate change.

To this end, education must succeed in bridging the gap between attitudes and actions; a child must understand and then think about what he or she can do. Without exposing the youths to what they do not know about climate change, and without highlighting how these global issues are affecting their lives at a local level, there is little motivation for these custodians of our environmental future to acquire sufficient knowledge, skills and understanding of a topic that is profoundly and inextricably woven into the fabric of our current environmental reality.

Bibliography

Abalahin, M., & Chang, C. H. (2020). Sustainability in primary geography. In C. Chang, G. Kidman & A. Wi (Eds.), *Issues in teaching and learning of education for sustainability theory into practice* (1 st ed., pp. 17–31). Singapore: Routledge.

Ahmad, A., Rahim, S., Pawanteh, L., & Ahmad, F. (2012). The understanding of environmental citizenship among Malaysian youths: A study on perception and participation. *Asian Social Science*, 8(5), 85–92.

Aikens, K., McKenzie, M., & Vaughter, P. (2016). Environmental and sustainability education policy research: A systematic review of methodological and thematic trends. *Environmental Education Research*, 22(3), 333–359.

Aldrich-Moodie, B., & Kwong, J. (1997). *Environmental education*. London: Institute of Economic Affairs.

Anderson, A. (2010). *Combating climate change through quality education*. Washington, DC: Brookings Global Economy and Development.

Andersson, B., & Wallin, A. (2000). Students' understanding of the greenhouse effect, the societal consequences of Reducing CO_2 emissions and the problem of ozone layer depletion. *Journal of Research in Science Teaching*, 37(10), 1096–1111.

Ariasi, N., & Mason, L. (2011). Uncovering the effect of text structure in learning from a science text: An eye-tracking study. *Instructional Science*, 39(5), 581–601. https://doi.org/10.1007/s11251-010-9142-5

Armstrong, H. W., & Read, R. (1998). Trade and growth in small states: The impact of global trade liberalisation. *World Economy*, 21(4), 563–585.

Arnold, W. (2007, Aug 29). Vulnerable to rising seas, Singapore envisions a giant seawall. Retrieved from *New York Times*: http://www.nytimes.com/2007/08/29/world/asia/29iht-Dikes.2.7301576.html?_r=0

Arrhenius, S. (1896). On the influence of carbonic acid in the air on the temperature of the ground. *Philosophical Magazine*, 41(251), 236–276.

Arslan, H. O., Cigdemoglu, C., Moseley, C. (2012). A three-tier diagnostic test to assess pre-service teachers' misconceptions about global warming, greenhouse effect, ozone layer depletion, and acid rain. *International Journal of Science Education*, 34(11), 1667–1686. http://doi.org/10.1080/09500693.2012.680618

Asia Pacific Energy Research Center. (2009). *APEC Energy Demand and Supply Outlook 4th Edition Economy Review*. Retrieved May 20, 2013, from Asia Pacific Energy Research Center: publications.apec.org/file-download.php?filename...Outlook.

Atasoy, B., Akkus, H., & Kadayifci, H. (2009). The effect of a conceptual change approach on understanding of students' chemical equilibrium concepts. *Research in Science & Technological Education*, 27(3), 267–282. https://doi.org/10.1080/02635140903162587

Bahrawi, N. (2007, June 2). PM Lee waves red flag over green issues. *Weekend Today*.

Baker, A. (2009, Feb 4). Don's urban farm idea wins accolade. *The Straits Times*, p. B5.

Balım, A. (2013). The effect of mind-mapping applications on upper primary students' success and inquiry-learning skills in science and environment education. *International Research in Geographical And Environmental Education*, 22(4), 337–352. https://doi.org/10.1080/10382046.2013.826543

Ballantyne, R., Connell, S., & Fien, J. (2006). Students as catalysts of environmental change: A framework for researching intergenerational influence through environmental education. *Environmental Education Research*, 12(3), 413–427.

Bangay, C., & Blum, N. (2010). Education responses to climate change and quality: Two parts of the same agenda? *International Journal of Educational Development*, 30(4), 359–368.

Barraza, L., Duque-Aristizabel, A., & Rebolledo, G. (2003). Environmental education: From policy to practice. *Environmental Education Research*, 9(3), 348–357.

Bayer. (2010). *To experience environmental protection first-hand*. Retrieved Aug 26, 2011, from Bayer Young Environmental Envoy: http://www.bayeryoungenvoy.com/en/default.aspx

Bengtsson, S. L., & Östman, L. O. (2016). Globalisation and education for sustainable development: exploring the global in motion. *Environmental Education Research*, *22*(1), 1–20.

Birdsall, S. (2004). Overcoming Scale When Teaching about the World's Complexity. *Journal of Geography*, 103(2), 47–54. https://doi.org/10.1080/00221340408978575

Black, P. (2000). Research and the development of educational assessment. *Oxford Review of Education*, 26, 37–41.

Block, T., Goeminne, G., & Van Poeck, K. (2018). Balancing the urgency and wickedness of sustainability challenges: Three maxims for post-normal education. *Environmental Education Research*, 24(9), 1424–1439. https://doi.org/10.1080/13504622.2018.1509302

Bloom, B. (1956). *Taxonomy of educational objectives, the classification of educational goals – Handbook I: Cognitive domain* (B. Bloom, Ed.) New York: McKay.

Bodzin, A. M., Anastasio, D., Sahagian, D., Peffer, T., Dempsey, C., & Steelman, R. (2014). Investigating climate change understandings of urban middle-level students. *Journal of Geoscience Education*, 62(3), 417–430. http://doi.org/10.5408/13-042.1

Bonnett, M. (2002). Education for sustainability as a frame of mind. *Environmental Education Research*, 8, 9–20.

Boon, H. (2010). Climate Change? Who Knows? A comparison of Secondary Students and Pre-service teachers. *Australian Journal of Teacher Education*, 35(1), 104–120.

Bord, R., O'Connor, R., & Fisher, A. (2000). In what sense does the public need to understand global climate change? *Public Understanding of Science*, 9(3), 205–218.

Boussalis, C., & Coan, T. (2016). Text-mining the signals of climate change doubt. *Global Environmental Change*, 36, 89–100. http://doi.org/10.1016/j.gloenvcha.2015.12.001

Boyer, L., & Roth, W. (2006). Learning and teaching as emergent features of informal settings: An ethnographic study in an environmental action group. *Science Education*, 90, 1028–1049.

Boyes, E., Chuckran, D., & Stanisstreet, M. (1993). How do high school students perceive global climatic change: What are its manifestations? What are its origins? What corrective action can be taken? *Journal of Science Education and Technology* 2, 541–557. https://doi.org/10.1007/BF00695323

Boyes, E., & Stanisstreet, M. (1993). The 'Greenhouse Effect': Children's perceptions of causes, consequences and cures. *International Journal of Science Education*, 15(5), 531–552.

Boyes, E., Stanisstreet, M., & Yongling, Z. (2008). Combating global warming: The ideas of high school students in the growing economy of south east China. *International Journal of Environmental Studies*, 65(2), 233–245.

Boylan, C. (2008). Exploring elementary students' understanding of energy and climate change. *International Electronic Journal of Elementary Education*, *1*(1), 1–15.

Bozdogan, A. E. (2011). The effects of instruction with visual materials on the development of preservice elementary teachers' knowledge and attitude towards global warming. *Turkish Online Journal of Educational Technology*, 10(2), 218–233.

Braten, I., Stromso, H., & Salmeron, L. (2011). Trust and mistrust when students read multiple information sources about climate change. *Learning and Instruction*, 21(2), 180–192.

Broughton, S. H., Sinatra, G. M., & Reynolds, R. E. (2010). The nature of the refutation text effect: An investigation of attention allocation. *The Journal of Educational Research*, 103(6), 407–423. https://doi.org/10.1080/00220670903383101

Bulkeley, H. (2000), "Common knowledge? Public understanding of climate change in Newcastle, Australia", *Public Understanding of Science*, 9(1), 313–333.

Burby, R. J. (2006). Hurricane Katrina and the paradoxes of government disaster policy: Bringing about wise governmental decisions for hazardous areas. *The Annals of the American Academy of Political and Social Science*, 604(1), 171–191.

Busch, T. & Judick, L. (2021). Climate change—that is not real! A comparative analysis of climate-sceptic think tanks in the USA and Germany. *Climatic Change*, 164, 18. https://doi.org/10.1007/s10584-021-02962-z

Caiman, C., & Lundegård, I. (2014). Preschool children's agency in learning for sustainable development. *Environmental Education Research*, 20(4), 437–459. http://doi.org/10.1080/13504622.2013.812722

Cajkler, W., Wood, P., Norton, J., & Pedder, D. (2014). Lesson study as a vehicle for collaborative teacher learning in a secondary school. *Professional Development in Education*, 40(4), 511–529. http://doi.org/10.1080/19415257.2013.866975

Cardullo, P., & Kitchin, R. (2019) Smart urbanism and smart citizenship: The neoliberal logic of 'citizen-focused' smart cities in Europe. *Environment and Planning C: Politics and Space*, 37(5), 813–830.

Casey, E. S. (1993). Getting back into place: Toward a renewed understanding of the place-world. *Politics and Space*, 37(5), 813–830. https://doi.org/10.1177/0263774X18806508

Catling, S. (2005) Children's personal geographies and the English primary school geography curriculum. *Children's Geographies*, 3(3), 325–344. http://doi.org/10.1080/14733280500353019

Cecioni, E. (2005). Environmental education and geography of complexity. *International Research in Geographical and Environmental Education*, 14(4), 277–294. https://doi.org/10.1080/10382040508668360

CEED Consulting. (2011). *Knowledge, attitudes and practices survey on children and climate change*. Retrieved July 19, 2013, from UN in Montenegro.

Chambers, B. (1999). Environmental education and primary geography. *International Research in Geographical and Environmental Education*, 8(1), 90–93.

Chang, C. H. (2008). *Climate and Climate Change: A Singapore Perspective*. Singapore: McGraw Hill.

Chang, C. H. (2010). The impact of Global Warming on storms and storm preparedness in Southeast Asia. *Kajian Malaysia*, 28(1), 53–82.

Chang, C. H. (2011). Preparedness and storm hazards in a global warming world: Lessons from Southeast Asia. *Natural Hazards*, 56(3), 667–679.

Chang, C. H. (2012a). A critical discourse of Singapore's school geography for the twenty-first century. *Literacy Information and Computer Education Journal*, 3(3), 717–729.

Chang, C. H. (2012b). *Climate Change Education: Perception and practices*. National Institute of Education, Office of Graduate Studies and Professional Learning. Singapore: National Institute of Education, Nanyang Technological University.

Chang, C. H. (2012c). The changing climate of teaching and learning school geography–The case of Singapore. *International Research in Geographical and Environmental Education*, 21(4), 1–13.

Chang, C. H. (2013). Advancing a framework for climate change education in singapore through teacher professional development. *HSSE Online Research and Practice in Humanities & Social Studies Education*, 2(1), 28–35.

Chang, C. H. (2014). *Climate change education: Knowing, doing and being*. Abingdon: Routledge.

Chang, C. H. (2015). Teaching climate change – A fad or a necessity? *International Research in Geographical and Environmental Education*, 24(3), 181–183. https://doi.org/10.1080/10382046.2015.1043763

Chang, C. H., & Goh, K. (2011). *A critical discourse on climate change education in Singapore*. Nanyang Technological University.

Chang, C. H. & Pascua, L. (2014). Uncovering the nexus between scientific discourse and school geography in Singapore students' understanding of climate change. *Research in Geographic Education*, 16(1), 41–56.

Chang, C. H., & Pascua, L. (2015). 'The hole in the sky causes global warming': A case study of secondary school students' climate change alternative conceptions. *Review of International Geographical Education Online*, 5(3), 316–331.

Chang, C. H., & Pascua, L. (2016). Singapore students' misconceptions of climate change. *International Research in Geographical and Environmental Education, 25*(1), 84–96.

Chang, C. H., & Pascua, L. (2017). The curriculum of climate change education: A case for Singapore. *The Journal of Environmental Education*, 48(3), 172–181. http://doi.org/10.1080/00958964.2017.1289883

Chang, C. H., & Pascua, L. (2018). Climate change and sustainability education: Singapore teachers' perspectives. In M. Merrill, P. Burkhardt-Holm, C. Chang, M. Islam & Y. Chang, *Education and sustainability: Paradigms, policies and practices in Asia*. Oxon: Routledge.

Chang, C. H., Pascua, L., & Ess, F. (2018). Closing the "Hole in the Sky": The use of refutation-oriented instruction to correct students' climate change misconceptions. *Journal Of Geography*, 117(1), 3–16. http://doi.org/10.1080/00221341.2017.1287768

Chang, C. H., Tan, G. C. I., Tan, J., & Kwek, C. H. (2020). Developing a learning progression for climate change in geography education. Technical Report, Office of Education Research. Singapore: National Institute of Education.

Changi Airport Group (2011) Earth hour takes off at Changi Airport for the third consecutive year. Retrieved June 1, 2013, from http://www.changiairport.com/our-business/airport-news/earth-hour-takes-off-at-changi-airport

Chatterjea, K. (2009). Severe wet spells and vulnerability of urban slopes: Case of Singapore. *Natural Hazards*, 56(1), 1–18.

Cheam, J. (2010) Stepping out of our comfort zones. Retrieved Aug 26, 2011, from http://blogs.straitstimes.com/2010/12/17/stepping-out-of-our-comfort-zones/

Cheong, I. P. A., Treagust, D., Kyeleve, I. J., & Oh, P. Y. (2010). Evaluation of students' conceptual understanding of malaria. *International Journal of Science Education*, 32(18), 2497–2519.

Chew, V. (2008, Oct 20). *Singapore Green Plan (SGP)*. Retrieved Feb 5, 2009, from Singapore Infopedia An electronic encyclopedia on Singapore's history, culture, people and events: http://infopedia.nl.sg/articles/SIP_1370_2008-11-22.html

Chi, M., & Roscoe, R. (2002). The process and challenges in conceptual change. In M. Limon & L. Mason (Eds.), *Reconsidering conceptual change: Issues in theory and practice* (pp. 3–27). Amsterdam: Kluwer Academic Publishers.

Chi, M. T. (2008). Three types of conceptual change: Belief revision, mental model transformation, and categorical shift. In Stella Vosniadou (Eds.), *International handbook of research on conceptual change* (pp. 61–82).

Chi, M. T. (2013). Two kinds and four sub-types of misconceived knowledge, ways to change it, and the learning outcomes. In Stella Vosniadou (Eds.), *The international handbook of conceptual change* (2nd ed., pp. 49–70). New York: Routledge.

Chia, L., Khan, H., & Chou, L. (1988). *The coastal environmental profile of Singapore*. Association of Southeast Asian Nations & United States Resource Management Project. Manila: International Center for Living Aquatic Resources Management.

Chib, A., Chiew, H., Kumar, C., Lim, G., & Ale, K. (2009). Influencing pro–environmental attitudes among Singaporean youth. *Environmental Education Research*, 15(6), 679–696.

Chicago Public School. (n.d.). *Selecting tasks for performance assessment.* Retrieved 2007, from Chicago Public School: http://intranet.cps.k12.il.us/Assessments/Ideas_and_Rubrics/Assessment_Tasks/Select_Task/select_task.html

Choo, E. (2010, Dec 23). *Singapore braces for climate change.* Retrieved Dec 28, 2010, from Channel NewsAsia: http://www.channelnewsasia.com/stories/singaporelocalnews/view/1100833/1/.html

Chua, B. (2007). Singapore economy: The new and the dual. In T. H. Tan (Ed.), *Singapore perspectives 2007: A new Singapore.* Singapur: Institute of Policy Studies.

Chua, T. (2010, Oct 27). *Study shows Singaporeans more concerned about economy than climate change.* Retrieved June 16, 2011, from Singapore Business Review: http://sbr.com.sg/ecnomy/in-focus/study-shows-singaporeans-more-concerned-about-economy-climate-change

Çimer, S. O., Çimer, A., Ursava, A. (2011). Student teachers' conceptions about global warming and changes in their conceptions during pre-service education: A cross sectional study. *Educational Research and Reviews*, 6(8), 592–597.

Clean & Green Singapore. (2010). Retrieved June 6, 2022, from https://www.cgs.gov.sg

Coca-Cola Singapore (2011) Local commitments. Retrieved June 1, 2013, from http://www.coca-cola.com.sg/climate/local_commitments.asp

Cohen, P. (1995). Designing performance assessment tasks. *Education Update*, 37(6), 1–5.

Cordero, E., Todd, A., & Abellera, A. (2008). Climate change education and the ecological footprint. *Bulletin of the American Meteorological Society*, 89(6), 865–872.

Corney, G. (2000). Student geography teachers' pre-conceptions about teaching environmental topics. *Environmental Education Research*, 6(4), 313–329.

Corney, G., & Reid, A. (2007). Student teachers' learning about subject matter and pedagogy in education for sustainable development. *Environmental Education Research*, 13(1), 33–54.

Coşkun, M., & Aydın, F. (2011). Geography teacher candidates' perceptions towards the "greenhouse effect". *American–Eurasian Journal of Agriculture & Environment Sciences*, 10(2), 290–295.

Cotton, D. R. (2006). Teaching controversial environmental issues: Neutrality and balance in the reality of the classroom. *Educational Research*, 48(2), 223–241. https://doi.org/10.1080/00131880600732306

Crooks, T. J., Kane, M. T., & Cohen, A. S. (1996). Threats to the valid use of assessments. *Assessment in Education: Principles, Policy & Practice*, 3(3), 265–286. http://doi.org/10.1080/0969594960030302

Cross, K. (1998). Classroom research: Implementing the scholarship of teaching. *New Directions for Teaching And Learning*, 1998(75), 5–12. http://doi.org/10.1002/tl.7501

CSRSingapore. (2009). *Project E.A.R.T.H. by NEA*. Retrieved Jan 1, 2013, from Singapore Compact: www.csrsingapore.org/c/news/46-earth

Curriculum Planning & Development Division. (2005a). *Social studies syllabus: Primary*. Retrieved Dec 28, 2010, from Curriculum Planning and Development Division, Ministry of Education Singapore: http://www.moe.gov.sg

Curriculum Planning & Development Division. (2005b). *Geography syllabus: Lower secondary*. Retrieved July 17, 2012, from Ministry of Education Singapore: http://www.moe.gov.sg/education/syllabuses/humanities/files/geography-lower-secondary-2006.pdf

Curriculum Planning & Development Division. (2006a). *Social studies: Secondary 2 Normal (Technical)*. Retrieved Dec 28, 2010, from Curriculum Planning and Development Division, Ministry of Education Singapore: http://www.moe.gov.sg

Curriculum Planning & Development Division. (2006b). *Social studies: Secondary 2 normal (Technical)*. Retrieved 2012, from Curriculum Planning and Development Division, Ministry of Education Singapore: http://www.moe.gov.sg

Curriculum Planning & Development Division. (2007a). *Science syllabus: Lower secondary: Express/Normal (Academic)*. Retrieved Dec 28, 2010, from Curriculum Planning and Development Division, Ministry of Education Singapore: http://www.moe.gov.sg

Curriculum Planning & Development Division. (2007b). *Civics and moral education syllabus: Primary 2007*. Retrieved 2013, from Curriculum Planning and Development Division, Ministry of Education, Singapore: http://www.moe.gov.sg

Curriculum Planning & Development Division. (2007c). *Science syllabus: Lower secondary: Express/Normal (Academic)*. Retrieved December 28, 2010, from http://www.moe.gov.sg Curriculum Planning and Development Division, Ministry of Education, Singapore.

Curriculum Planning & Development Division. (2013). Science syllabus: Lower secondary: Express/Normal (Academic). Retrieved December 28, 2013, from Curriculum Planning and Development Division, Ministry of Education Singapore. http://www.moe.gov.sg

Curriculum Planning and Development Division. (2016a). Combined Humanities syllabus upper secondary. Singapore: Ministry of Education. Retrieved 2017, from Curriculum Planning and Development Division, Ministry of Education, Singapore: http://www.moe.gov.sg

Curriculum Planning and Development Division. (2016b). *Geography syllabus pre-university H1*. Singapore: Ministry of Education.

Curriculum Planning and Development Division. (2016c). *Geography syllabus pre-university H2* (p. 8). Singapore: Ministry of Education.

Curriculum Planning and Development Division. (2017). *Geography syllabus upper secondary*. Singapore: Ministry of Education.

Curriculum Planning and Development Division. (2020). *Biology syllabus higher 1* (p. 24). Singapore: Ministry of Education.

Curriculum Planning and Department Division. (2021a). *Geography teaching and learning syllabuses lower secondary* (p. 9). Singapore: Ministry of Education.

Curriculum Planning and Department Division. (2021b). *Science syllabuses lower secondary* (pp. 34–52). Singapore: Ministry of Education.

Damico, & Baidon. (2011). Content literacy for the 21st century: Excavation and elevation and relational cosmopolitanism in the classroom. *Journal of Adolescent & Adult Literacy*, 55(3), 232–243.

Daniel, B., Stanisstret, M., & Boyes, E. (2007). How can we best reduce global warming? School students' ideas and misconceptions. *International Journal of Environmental Studies*, 61(2), 211–222.

Davidson, P. (2003). Deconstructing the four pillars of the climate change debate: A critical review of the scientific, economic, political and ethical dimensions. *FES Outstanding Graduate Student Paper Series*, 8(8), 1–67.

Dawson, V., & Carson, K. (2013). Australian secondary school students' understanding of climate change. *Teaching Science*, 59(3), 9–14.

Delors, J. (1998). *Learning: The treasure within*. Paris: UNESCO.

Department of Statistics. (2022a). *Climate and air quality*. Singapore: Government of Singapore. https://www.singstat.gov.sg/publications/reference/ebook/society/climate-and-air-quality

Department of Statistics. (2022b). *Enrolment in educational institutions*. Singapore: Government of Singapore. https://tablebuilder.singstat.gov.sg/table/TS/M850251

Department of Statistics Singapore. (2012). *Latest Data*. Retrieved July 22, 2013, from Department of Statistics Singapore: http://www.singstat.gov.sg/statistics/latest_data.html#12

Detenber, B., Rosenthal, S., Liao, Y., & Ho, S. (2016). Audience segmentation for campaign design: Addressing climate change in Singapore. *International Journal Of Communication*, 10, 4736–4758.

Dewey, J. (1997). *Experience and education*. New York: Simon & Schuster.

Doering, A., & Veletsianos, G. (2008). An investigation of the use of real-time, authentic geospatial data in the K–12 classroom. *Journal of Geography*, 106(6), 217–225. https://doi.org/10.1080/00221340701845219

Dudley, P. (2013). Teacher learning in lesson study: What interaction-level discourse analysis revealed about how teachers utilised imagination, tacit knowledge of teaching and fresh evidence of pupils learning, to develop practice knowledge and so enhance their pupils' lea. *Teaching and Teacher Education*, 34, 107–121. http://dx.doi.org/10.1016/j.tate.2013.04.006

Duit, R. (1999) Conceptual change approaches in science education. In W. Schnotz, S. Vosniadou, & M. Carretero (Eds.), *New perspectives on conceptual change* (pp. 263–282). Amsterdam: Pergamon.

Duit, R., & Treagust, D. F. (2003) Conceptual change: A powerful framework for improving science teaching and learning. *International Journal of Science Education*, 25, 671–688. http://dx.doi.org/10.1080/09500690305016

Dupigny-Giroux, L. A. L. (2010). Exploring the challenges of climate science literacy: Lessons from students, teachers and lifelong learners. *Geography Compass*, 4(9), 1203–1217.

Eilam, E. & Trop, T. (2010) ESD pedagogy: A guide for the perplexed. *The Journal of Environmental Education*, 42(1), 43–64. DOI:10.1080/00958961003674665

Elwood, S. (2004). Experiential learning, spatial practice, and critical urban geographies. *Journal of Geography*, 103(2), 55–63. https://doi.org/10.1080/00221340408978576

Fang, J. (2008, Sep 10). PM calls for US leadership on climate. *The Straits Times*, p. 3.

Feierabend, T., Jokmin, S., & Eilks, I. (2011). Chemistry teachers' views on teaching 'climate change' – An interview case study from research-orientated learning in teacher education. *Chemistry Education Research and Practice*, 12, 85–91.

Fernandez, C., & Yoshida, M. (2004). *Lesson study: A Japanese approach to improving mathematics teaching and learning.* Mahwah, NJ: Lawrence Erlbaum Associates.

Field, C. B. et al. (2014) Technical summary. In C. B. Field, V. R. Barros, D. J. Dokken, K. J. Mach, M. D. Mastrandrea, T. E. Bilir, M. Chatterjee, K. L. Ebi, Y. O. Estrada, R. C. Genova, B. Girma, E. S. Kissel, A. N. Levy, S. MacCracken, P. R. Mastrandrea, & L. L. White (Eds.), *Climate change 2014: Impacts, adaptation, and vulnerability. Part A: Global and sectoral aspects. Contribution of working group II to the fifth assessment report of the intergovernmental panel on climate change.* Cambridge: Cambridge University.

Fien, J. (1993). *Environmental education: A pathway to sustainability.* Geelong: Deakin University.

Fien, J., Heck, D., & Ferreira, J. (1997). *Learning for a sustainable environment.* Paris: UNESCO.

Figueres, C. (2012, Oct 18). Climate change: Why we need a multilateral solution. *S. Rajaratnam School of International Studies Lecture.* Singapore.

Firth, R., & Winter, C. (2007). Constructing education for sustainable development: The secondary school geography curriculum and initial teacher training. *Environmental Education Research*, 13(5), 599–619.

Fortner, R. (2001). Climate change in School: Where does it fit and how ready are we? *Canadian Journal of Environmental Education*, 6, 18–31.

Fourier, J. (1827). Memoire sur les temperatures du globe terrestre et des espace planetaires. *Memoires de l'academie Royal des Sciences de l'Institut de France* (VII), 7, 570–604.

Francis, C., Boyes, E., Qualter, A., & Stanisstreet, M. (1993). Ideas of elementary students about reducing the "Greenhouse Effect". *Science Education*, 77(4) 375–392.

Gagné, R. (1985). *The conditions of learning and theory of instruction* (4th ed.). New York: Holt, Rinehart & Winston.

Gagné, R., & Medsker, K. (1996). *The conditions of learning. Training applications* (1st ed.). Fort Worth, TX: Harcourt Brace College Pub.

Gagné, R. M., & Driscoll, M. P. (1988). *Essentials of teaching.* Englewood Cliffs, NJ: Prentice Hall.

Gautier, C., Deutsch, K., & Rebich, S. (2006). Misconceptions about the greenhouse effect. *Journal of Geoscience Education*, 54, 386.

Gilani Research Foundation. (2009, Feb 13). *Business and politics in the Muslim World.* Retrieved Aug 2, 2009, from Gilani Research Foundation: www.gilanifoundation.com/homepage/Collected_paper08-09/13-February%202009.pdf

Global Greenhouse Warming. (2010). *Climate mitigation and adaptation.* Retrieved Jan 10, 2012, from Global Greenhouse Warming: http://www.global-greenhouse-warming.com/climate-mitigation-and-adaptation.html

Goh, K., Tan, K., Chang, C., & Ooi, G. (2009). Climate change and sustainable development: The response from education – The case of Singapore. In J. Laessoe, S. Karsten, S. Breiting, & S. Rolls (Eds.), *The joint national reports on climate change and sustainable development: The response from education* (pp. 240–291). Denmark: The International Alliance of leading Education Institutes.

Goh-Ice, R. (2000). Environmental education in the United States: A survey of pre-service teacher education programs. *Journal of Environmental Education*, 32(1), 4–11.

Gold, A., Oonk, D., Smith, L., Boykoff, M., Osnes, B., & Sullivan, S. (2015). Lens on climate change: Making climate meaningful through student-produced videos. *Journal of Geography*, 114(6), 235–246. https://doi.org/10.1080/00221341.2015.1013974

Gowda, M. V., Fox, J., & Magelky, R. (1997). Students' understanding of climate change: Insights for scientists and educators. *Bulletin of the American Meteorological Society*, 78(10), 2232–2240.

Grace, M., & Sharp, J. (2000) Exploring the actual and potential rhetoric-reality gaps in environmental education and their implications for pre-service teacher training. *Environmental Education Research*, 6(4), 331–345. http://doi.org/10.1080/713664698

Grant, L., & Featherstone, H. (2009, July). *Climate change schools Project: Final evaluation report.* Science Learning Centre, UK. Retrieved June 16, 2011, from Climate Change Schools Project: http://www.sciencelearningcentres.org.uk/centres/north-east/cliamte-change-schools-project/CCSP_evaluation_final_report_July09.pdf

Griffin, P. & Care, E. (2015). Assessment and TEACHING of 21st century skills: Methods and approach. http://doi.org/10.1007/978-94-017-9395-7

Gunasingham, A. (2010, Mar 9). Study points to rise in Singapore temperature, sea levels. *The Straits Times.*

Guzzetti, B., Williams, W. O., Skeels, S. A., & Wu, S. M. (1997). Influence of text structure on learning counterintuitive physics concepts. *Journal of Research in Science Teaching*, 34(7), 701–719. https://doi.org/10.1002/(SICI)1098-2736(199709)34:7<701::AID-TEA3>3.0.CO;2-Q

Guzzetti, B. J. (2000). Learning counter-intuitive science concepts: What have we learned from over a decade of research? *Reading & Writing Quarterly: Overcoming Learning Difficulties*, 16(2), 89–98. https://doi.org/10.1080/105735600277971

Hagstrom, F. (2006). Formative learning and assessment. *Communication Disorders Quarterly*, 28, 24–36.

Hamilton-Hart, N. (2006). Singapore's climate change policy: The limits of learning. *Contemporary Southeast Asia*, 26(3), 363–384.

Hatano, G. & Inagaki, K. (1991). Sharing cognition through collective comprehension activity. In L. B. Resnick, J. M. Levine & S. D. Teasley (Ed.), *Perspectives on socially shared cognition* (pp. 331–348). Washington DC: American Psychological Association

Han, Q. (2015). Education for sustainable development and climate change education in China: A status report. *Journal of Education for Sustainable Development*, 9(1), 62–77. http://doi.org/10.1177/0973408215569114

Hansen, P. J. K. (2010). Knowledge about the greenhouse effect and the effects of the ozone layer among Norwegian pupils finishing compulsory education in 1989, 1993, and 2005- what now? *International Journal of Science Education*, 32(3), 397–419. http://doi.org/10.1080/09500690802600787

Harding, N. (2012, July 14). *Sir David Attenborough: 'This awful summer? We've only ourselves to blame…'*. Retrieved July 1, 2013, from The Independent: http://www.independent.co.uk/news/people/profiles/sir-david-attenborough-this-awful-summer-weve-only-ourselves-to-blame-7942405.html

Harrington, J. (2008). Misconceptions: Barriers to improved climate literacy. *Physical Geography*, 29(6), 575–584.

Haubrich, H. (2000). Guest editorial: Sustainable learning in geography for the 21st century. *International Research In Geographical And Environmental Education*, 9(4), 279–284. http://doi.org/10.1080/10382040008667660

Haubrich, H., Reinfried, S. Y., & Schleicher, Y. (2007). Lucerne declaration on geographical education for sustainable development. In: S. Reinfried, Y. Schleicher, & A. Rempfler (Eds.), *Geographical views on Education for sustainable development. Proceedings of the Lucerne-Symposium, Geographiedidaktische forschungen*. Lucerna: IGU-UGI.

Heimlich, J., Braus, J., Olivolo, B., McKeown-Ice, R., & Barringer-Smith, L. (2004). Environmental education and preservice teacher preparation: A national study. *Journal of Environmental Education*, 35(2), 17–60.

Henderson, J. C. (2012). Planning for success: Singapore, the model city-state? *Journal of International Affairs*, 65(2), 69–83.

Henry, A. (2000). Public perceptions of global warming. *Research in Human Ecology*, 7(1), 25–30.

Hensley, N. (2011). *Curriculum studies gone wild: Bioregional education and the scholarship of sustainability*. New York: Peter Lang.

Hermans, M., & Korhonen, J. (2017). Ninth graders and climate change: Attitudes towards consequences, views on mitigation, and predictors of willingness to act, *International Research in Geographical and Environmental Education*, 26(3), 223–239.

Hestness, E., McGinnis, J., Riedinger, K., & Marbach-Ad, G. (2011). A study of teacher candidates' experiences investigating global climate change within an elementary science methods course. *Journal of Science Teacher Education*, 22(4), 351–369.

Heywoth, R. (2008). *New science discovery for lower secondary* (2nd ed., Vol. 2). Singapore: Pearson Education.

Higgit, D. (2006). Finding space for education for sustainable development in the enterprise economy. *Journal of Geography in Higher Education*, 30(2), 251–262.

Ho, L. & Seow, T. (2015) Teaching controversial issues in geography: Climate change education in Singaporean schools. *Theory & Research in Social Education*, 43(3), 314–344. http://doi.org/10.1080/00933104.2015.1064842

Ho, L., & Seow, T. (2017). Disciplinary boundaries and climate change education: Teachers' conceptions of climate change education in the Philippines and Singapore. *International Research In Geographical And Environmental Education*, 26(3), 240–252. https://doi.org/10.1080/10382046.2017.1330038

Ho, S., & Chuah, A. (2017). Climate change communication in Singapore. Oxford Research Encyclopedia Of Climate Science (pp. 500–514). http://doi.org/10.1093/acrefore/9780190228620.013.473

Hongkong and Shanghai Banking Corporation. (2007). *HSBC commits to change: Case studies.* Retrieved Aug 2, 2008, from Hongkong and Shanghai Banking Corporation: http://www.hsbccommittochange.com/environment/hsbc-case-studies/index.aspx

Huckle, J. (1985). Geography and schooling. In R. J. Johnston (ed.), *The Future of Geography* (pp. 291–306). Methuen.

Hussain, Z. (2007, Apr 23). Government preparing for possibility of rising sea levels. *The Straits Times.*

Hynd, C. R. (2001). Refutational texts and the change process. *International Journal of Educational Research*, 35(7–8), 699–714.

Ikonomidis, S., Papanastasiou, D., Melas, D., Avgoloupis, S. (2012). The anthropogenic 'greenhouse effect': Greek prospective primary teachers' ideas about causes, consequences and cures. *Journal of Science Education and Technology*, 21(6), 768–779. http://doi.org/10.1007/s10956-012-9365-0

Inagaki, K., & Hatano, G. (2003). Conceptual and linguistic factors in inductive projection: How do young children recognize commonalities between. In D. Gentner and S. Goldin-Meadow (Eds.), *Language in mind: Advances in the study of language and thought* (pp. 313–334). Cambridge: MIT Press.

Intergovernmental Panel on Climate Change. (2007). *Fourth assessment report: Climate change 2007: The AR4 synthesis report.* Intergovernmental Panel on Climate Change. Geneva: IPCC.

Intergovernmental Panel on Climate Change. (2014). *Climate change 2014: Synthesis report. Contribution of working groups I, II and III to the fifth assessment report of the intergovernmental panel on climate change.* Intergovernmental Panel on Climate Change. Geneva: IPCC.

International Geographical Union Commission on Geographical Education (IGU-CGE). (1992). *International charter on geographical education.* Freiburg: IGU-CGE.

International Geographical Union Commission on Geographical Education (IGU-CGE). (2015). *The 2016 International charter on geography education: Draft June 10, 2015.* Beijing: IGU-CGE.

International Geographical Union - Commission on Geographical Education. (2016). International Charter on Geographical Education. Retrieved August 2016, from International Geographic Union - Commission on Geographical Education: http://www.igu-cge.org/Charters-pdf/2016/IGU_2016_def.pdf

Chang, C. H., & Pascua, L. (2016). Singapore students' misconceptions of climate change. *International Research in Geographical and Environmental Change*, 25(1), 84–96.

Jackson, L., & Pang, M. F. (2017). Secondary school students' views of climate change in Hong Kong, *International Research in Geographical and Environmental Education*, 26(3), 180–192.

Jeffries, H., Stanisstreet, M., & Boyes, E. (2001). Knowledge about the 'Greenhouse Effect': Have college students improved? *Research in Science & Technological Education*, 19(2), 205–211.

Jickling, B. (1992) Viewpoint: Why I don't want my children to be educated for sustainable development. *The Journal of Environmental Education*, 23(4), 5–8. http://doi.org/10.1080/00958964.1992.9942801

Jiuan, T. S., Wirtz, J., Junk, K., & Keng, K. A. (2001). Singaporeans' attitudes towards work, pecuniary adherence, materialism, feminism, environmental consciousness, and media credibility. *Singapore Management Review*, 23(1), 59–86.

Johnson, R. (2011). *Climate change education in K-12: Teacher preparation, understanding, needs and concerns.* Retrieved July 16, 2012, from National Academies ORG: http://www.nationalacademies.org/bose/1CCEW2_Roberta_Johnson.pdf

Jucker, R. (2011). ESD between systemic change and bureaucratic obfuscation: Some reflections on environmental education and education for sustainable development in Switzerland. *Journal of Education for Sustainable Development*, 5(1), 39–60. https://doi.org/10.1177/097340821000500109

Kagawa, F., & Selby, D. (2012). Ready for the storm: Education for disaster risk reduction and climate change adaptation and mitigation. *Journal of Education for Sustainable Development*, 6(2), 207–217.

Kastrenakes, J. (2014). States are fighting new science curriculum teaching climate change and evolution. *The Verge*. Retrieved 15 June 2022 from http://www.theverge.com/2014/5/19/5730600/states-debate-ngss-science-standards-teaching-climate-change-evolution

Kendeou, P., & Van Den Broek, P. (2007). The effects of prior knowledge and text structure on comprehension processes during reading of scientific texts. *Memory & Cognition*, 35(7), 1567–1577.

Kenis, A., & Mathijs, E. (2012). Beyond individual behaviour change: The role of power, knowledge and strategy in tackling climate change. *Environmental Education Research*, 18(1), 45–65.

Khalid, T. (2003). Pre-service high school teachers' perceptions of three environmental phenomena. *Environmental Education Research*, 9(1), 35–50.

Kilinc, A., Stanisstreet, M., & Boyes, E. (2008). Turkish students' ideas about global warming. *International Journal of Environmental and Science Education*, 3(2), 89–98.

Kirman, J. (2003). Transformative geography: Ethics and action in elementary and secondary geography education. *Journal of Geography*, 102(3), 93–98. https://doi.org/10.1080/00221340308978530

Koehler, M. J., & Mishra, P. (2009). What is technological pedagogical content knowledge? *Contemporary Issues in Technology and Teacher Education*, 9(1). Retrieved from http://www.citejournal.org/vol9/iss1/general/article1.cfm

Kolleck, N., & Yemini, M. (2020). Environment-related education topics within global citizenship education scholarship focused on teachers: A natural language processing analysis. *The Journal of Environmental Education*, 51(4), 317–331. https://doi.org/10.1080/00958964.2020.1724853

Kong, L., Teh-Cheong, I., Gusti Tina, P., Remorin, P., Suwannatachote, R., & Lee, W. (2000). Unity and diversity: South East Asia. In D. Yencken, J. Fien, H. Sykes, D. Yencken, J. Fien, & H. Sykes (Eds.), *Environment, Education and Society in the Asia-Pacific: Local traditions and global discourses.* London: Routledge.

Kong, L., Yuen, B., Sodhi, N. S., & Briffett, C. (1999). The construction and experience of nature: Perspectives of urban youths. *Tijdschrift Voor Economische En Sociale Geografie*, 90(1), 3–16.

Koo, E. (2007, Apr 25). World is in 'serious trouble' over global warming: MM. *The Stratis Times*, p. 3.

Korson, C., & Kusek, W. (2016). The comparison of a thematic versus regional approach to teaching a world geography course. *Journal of Geography*, 115(4), 159–168. https://doi.org/10.1080/00221341.2015.1076498

Koulaidis, V., & Christidou, V. (1999). Models of students' thinking concerning the greenhouse effect and teaching implications. *Science Education*, 83(5), 559–576.

Krasny, M., Lundholm, C., & Plummer, R. (2010). Environmental education, resilience, and learning: Reflection and moving forward. *Environmental Education Research*, 16(5–6), 665–672. https://doi.org/10.1080/13504622.2010.505445

Krathwohl, D. (2002). A revision of bloom's taxonomy: An overview. *Theory into Practice*, 41(4), 212–218.

Kuo, S. Y. (2010). *The Public Understanding of Climate Change: A Case Study of Taiwanese Youth*. Faculty of New Jersey Institute of Technology and Rutgers, The State University of New Jersey-Newark. A Dissertation Submitted in Partial Fulfillment of the Requirements for the Degree of Doctor of Philosophy in Environmental Science Department of Chemistry and Environmental Science.

Læssøe, J., & Mochizuki, Y. (2015). Recent trends in national policy on education for sustainable development and climate change education. *Journal of Education for Sustainable Development*, 9(1), 27–43.

Lambert, D., & Hopkin, J. (2014). A possibilist analysis of the national curriculum in England. *International Research in Geographical and Environmental Education*, 23(1), 64–78.

Lambert, J. L., Lindgren, J., & Bleicher, R. (2012). Assessing elementary science methods students' understanding about global climate change. *International Journal of Science Education*, 34(8), 1167–1187. http://doi.org/10.1080/09500693.2011.633938

Larkin, D. (2012). Misconceptions about "misconceptions": Preservice secondary science teachers' views on the value and role of student ideas. *Science Education*, 96(5), 927–959. http://doi.org/10.1002/sce.21022

Lau, F. (2006, Nov 6). PM's call: Make Singapore a city of gardens and water. *The Straits Times*, p. 6.

Lawson, V. (2007). Geographies of care and responsibility. *Annals of the Association of American Geographers*, 97(1), 1–11. https://doi.org/10.1111/j.1467-8306.2007.00520.x

Leal Filho, W., Pace, P., & Manolas, E. (2010). The contribution of education towards meeting the challenges of climate change. *Journal of Baltic Science Education*, 9(2), 142–155.

Lee, H. L. (2019). National Day Rally 2019. Retrieved Mar 16, 2022, from https://www.pmo.gov.sg/Newsroom/National-Day-Rally-2019

Lee, O., Lester, B., Ma, L., Lambert, J., & Jean-Baptiste, M. (2007). Conceptions of the greenhouse effect and global warming among elementary students from diverse languages and cultures. *Journal of Geoscience Education*, 55(2), 117–125.

Leiserowitz, A., Smith, N., & Marlon, J. (2011). *American Teens' Knowledge of Climate Change*. Retrieved June 16, 2011, from Yale Project on Climate Change Communication: http://environment.yale.edu/uploads/american-teens-knowledge-of-climate-change.pdf

Lesgold, A. (2001). The nature and methods of learning by doing. *American Psychologist*, 56(11), 964–973.

Lewis, C., Perry, R., & Murata, A. (2006). How should research contribute to instructional improvement? The case of lesson study. *Educational Researcher*, 35(3), 3–14.

Li, S., & Cecilia, L. (2011). Climate change and morality: Students' perspectives on the individual and society. *International Journal of Science Education*, 33(8), 1131–1148.

Liarakou, G., Athanasiadis, I., & Gavrilakis, C. (2011). What Greek secondary school students believe about climate change? *International Journal of Environmental and Science Education*, 6(1), 79–98.

Lim, E. Q. (2018). The role of geographical investigations in developing students' cognitive thinking. *HSSE Online*, 7(10), 14–26. http://dx.doi.org/10.1016/S0959-4752(00)00037-2

Lim, H. B. (2011). The national water study project. Retrieved Aug 26, 2011, from http://nwsp.ntu.edu.sg/nwsp/

Limón, M. (2001) On the cognitive conflict as an instructional strategy for conceptual change: A critical appraisal. *Learning and Instruction*, 11, 357–380.

Lin, C. (2021, Mar 4). Sustainability targets for built environment sector under newest Green Building Masterplan. Retrieved Mar 16, 2022, from https://www.channelnewsasia.com/singapore/green-building-masterplan-new-targets-sustainability-bca-1883121

Linn, R. (2005). Evaluating the validity of assessments: The consequences of use. *Educational Measurement: Issues and Practice*, 16(2), 14–16. http://doi.org/10.1111/j.1745-3992.1997.tb00587.x

Lowe, T. (2006). *Vicarious Experience Vs. l Scientific Information in Climate Change Risk Perception and Behaviour: A Case Study of Undergraduate Students in Norwich, UK*. Technical Report 43, Tyndall Centre for Climate Change Research, UK.

Luber, L., Fögele, J., Mehren, R. (2020). How do students experience a deprived urban area in Berlin? Empirical reconstruction of students' orientations. *Review of International Geographical Education (RIGEO)*, 10(4), 500–533. http://doi.org/10.33403.rigeo.763170

Marginson, S., & Dang, T. (2016). Vygotsky's sociocultural theory in the context of globalization. *Asia Pacific Journal of Education*, 37(1), 116–129. http://doi.org/10.1080/02188791.2016.1216827

Marginson, S., & Rhoades, G. (2002). Beyond national states, markets, and systems of higher education: A glonacal agency heuristic. *Higher Education*, 43, 281–309. https://doi.org/10.1023/A:1014699605875

Marsden, B. (1997). Continuity after the national curriculum. *Teaching Geography*, 68–70.

McCaffrey, M. S., & Buhr, S. M. (2008). Clarifying climate confusion: Addressing systemic holes, cognitive gaps, and misconceptions through climate literacy. *Physical Geography*, 29(6), 512–528.

Mason, L., & Santi, M. (1998). Discussing the greenhouse effect: Children's collaborative discourse reasoning and conceptual change. *Environmental Education Research*, 4(1), 67–85.

Matthes, F. E. (1939). Report of the committee on glaciers. *Transactions of the American Geophysical Union*, 20, 518–523.

McBean, G., & Hengeveld, H. (2000). Communicating the science of climate change: A mutual challenge for scientists and educators. *Canadian Journal of Environmental Education*, 5, 9–23.

McClaren, M. (2019). Revisioning environmental literacy in the context of a global information and communications ecosphere. *The Journal of Environmental Education*, 50(4–6), 416–435. https://doi.org/10.1080/00958964.2019.1687408

McCuin, J. L., Hayhoe, K., & Hayhoe, D. (2014). Comparing the effects of traditional vs. Misconceptions-based instruction on student understanding of the greenhouse effect. *Journal of Geoscience Education*, 62(3), 445–459. http://doi.org/10.5408/13-068.1

McKeown, R., & Hopkins, C. (2010). Rethinking climate change education. *Green Teacher*, 89, 17–21.

McKeown-Ice, R. (2000). Environmental education in the United States: A survey of preservice teacher education programs. *Journal of Environmental Education*, 32(1), 4–11.

McKie, R. (2009, Jan 18). *We have only four years left to act on climate change - America has to lead.* Retrieved July 1, 2013, from *The Guardian*: http://www.guardian.co.uk/environment/2009/jan/18/obama-climate-change

McNeill, K. L., & Vaughn, M. H. (2012). Urban high school students' critical science agency: Conceptual understandings and environmental actions around climate change. *Research in Science Education*, 42(2), 373–399.

McTighe, J., & Self, E. (2002). *Observable indicators of teaching or understanding.* TTL Academies. Alexandria, VA: Association for Supervision and Curriculum Development (ASCD.

Mehlsen, C. (2009, Sept). *Snapshots from a leading Eco-city.* Retrieved June 16, 2011, from Education Alliance Quarterly: http://www.google.com/url?sa=t&source=web&cd=1&ved=0CBMQFjAA&url=http%3A%2F%2Fwww.intlalliance.org%2Ffileadmin%2Fuser_upload%2Fdocuments%2Fialei_magazine%2Fsustainable_development%2Feducation_alliance_quarterly_vol.4_p22-23.pdf

Mehrens, W. A. (1997). The consequences of consequential validity. *Educational Measurement: Issues and Practices*, 16, 16–18. https://doi.org/10.1111/j.1745-3992.1997.tb00588.x

Meinhold, J. L., & Malkus, A. J. (2005). Adolescent environmental behaviors can knowledge, attitudes, and self-efficacy make a difference? *Environment and Behavior*, 37(4), 511–532.

Messick, S. (1995). Validity of psychological assessment: Validation of inferences from persons' responses and performances as scientific inquiry into score meaning. *American Psychologist*, 50(9), 741–749. https://doi.org/10.1037/0003-066X.50.9.741

Meyer, B. J. F. (2003). Text coherence and readability. *Topics in Language Disorders* 23, 204–224.

Mezirow, J. (1997). "Transformative learning: Theory to practice." *New Directions for Adult and Continuing Education*, 1997(74), 5–12.

Miller, J. (2012). Climate change: Generation X attitudes, interest and understanding. *The Generation X Report: A Quarterly Research Report from the Longitudinal Study of American Youth*, 1(4), 1–7.

Ministry of Education (2022). Education Statistics Digest. Retrieved 15 June 2022 from https://www.moe.gov.sg/about-us/publications/education-statistics-digest

Ministry of Education, Singapore. (2009). *Desired outcomes of Education.* Retrieved Apr 3, 2013, from Ministry of Education: http://www.moe.gov.sg/education/files/desired-outcomes-of-education.pdf

Ministry of Education, Singapore. (2012). *Education Statistics Digest 2012.* Retrieved July 22, 2013, from Ministry of Education: http://www.moe.gov.sg/education/education-statistics-digest/files/esd-2012.pdf

Ministry of Education, Singapore. (2012). *MOE Corporate Brochure: Singapore Education.* Retrieved July 22, 2013, from Ministry of Education: http://www.moe.gov.sg/about/files/moe-corporate-brochure.pdf

Ministry of Environment & Water Resources, Ministry of National Development. (2016). Singapore Sustainable Blueprint. Singapore: Centre of Liveable Cities, Singapore. Retrieved June 5, 2022, from Centre for Liveable Cities: https://www.clc.gov.sg/docs/default-source/books/ssbcombined-cover-text.pdf

Ministry of Environment and Water Resources, Singapore. (2006, Feb). *The Singapore Green Plan 2012, 2006 Edition.* Retrieved Aug 26, 2011, from Ministry of Environment and Water Resources: http://app.mewr.gov.sg/data/ImgCont/1342/sgp2012_2006edition.pdf

Ministry of Environment and Water Resources, Singapore. (2008, Mar). *Singapore's National Climate Change Strategy.* Retrieved Dec 28, 2010, from Singapore's National Climate Change Strategy: http://app.mewr.gov.sg/data/ImgUpd/NCCS_Full_Version.pdf

Ministry of Sustainability & Environment. (2021). Retrieved Dec 21, 2021, from https://www.mse.gov.sg/climate-action-week-2021/

Ministry of Sustainability & Environment. (2022). Retrieved Mar 16, 2022, from https://www.mse.gov.sg/resource-room/category/2021-08-26-press-release-on-second-grant-call-for-sg-eco-fund/

Mochizuki, Y., & Bryan, A. (2015). Climate change education in the context of education for sustainable development: Rationale and principles. *Journal of Education for Sustainable Development*, 9(1), 4–26. https://doi.org/10.1177/0973408215569109

Mohan, L. (2018). Student learning of place: Learning progressions can help fill the gaps. *Journal of Geography*, 117(3), 125–127. https://doi.org/10.1080/00221341.2017.1391313

Morgan, R. K. (2012) Environmental impact assessment: The state of the art. *Impact Assessment and Project Appraisal*, 30(1), 5–14. http://doi.org/10.1080/14615517.2012.661557

Mower, T. (2012). Climate change in the curriculum: Are all young people informed and inspired? *Earth & Environment*, 8(1), 1–37.

Muller, D., Bewes, J., Sharma, M., & Reimann, P. (2008). Saying the wrong thing: Improving learning with multimedia by including misconceptions. *Journal of Computer Assisted Learning*, 24, 144–155.

National Aeronautics and Space Administration. (2012). *Global mean land-ocean temperature index 1880 to present.* Retrieved 2013, from National Aeronautics and Space Administration Goddard Institute for Space Studies: http://data.giss.nasa.gov/gistemp/graphs/

National Climate Change Secretariat. (2011, Feb 24). *Info on climate change.* Retrieved July 17, 2012, from National Climate Change Secretariat: http://app.nccs.gov.sg/page.aspx?pageid=44

National Climate Change Secretariat. (2012). *National climate change strategy.* Retrieved July 17, 2012, from National Climate Change Secretariat: http://app.nccs.gov.sg/data/resources/docs/Documents/NCCS-2012.pdf

National Climate Change Secretariat. (2022). *National climate change strategy.* Retrieved May 1, 2022, from National Climate Change Secretariat: https://www.nccs.gov.sg/singapores-climate-action/singapore-emissions-profile/

National Environment Agency. (2002). *About the NEA*. Retrieved Aug 2, 2008, from National Environment Agency: http://app.nea.gov.sg/cms/htdocs/category_sub.asp?cid=2

National Environment Agency. (2006). *Awareness of environmental issues among students yet to be translated into actions.* Retrieved July 19, 2013, from National Environment Agency: http://app2.nea.gov.sg/news_detail_2006.aspx?news_sid=20081014961225796534

National Environment Agency. (2008). *NEA to households: Cut your energy bills by 10%.* Retrieved May 13, 2008, from National Environment Agency: http://app.nea.gov.sg/cms/htdocs/article.asp?pid=3025

National Environment Agency. (2010). Singapore's Second National Communication: Under the United Nations frame-work convention on climate change. Singapore: National Environment Agency. Retrieved 2020 from https://www.nea.gov.sg/docs/default-source/our-services/climate-change/second-nc_12-nov-2010.pdf

National Environment Agency. (2012). E^2 Singapore, Retrieved 2012 from https://policy.asiapacificenergy.org/sites/default/files/E2S%20Publication.pdf

National Environment Agency. (2013, June 21). *PSI reading.* Retrieved June 21, 2013, from National Environment Agency: www.nea.gov.sg/psi/

National Environment & Agency. (2020) *Waste Statistics and Overall Recycling.* Retrieved 24 August 2021, from https://www.nea.gov.sg/our-services/waste-management/waste-statistics-and-overall-recycling

National Environment Agency. (n.d.). *E2 Singapore.* Retrieved July 1, 2011, from National Environment Agency: www.nea.gov.sgcms/ccird/E2%20Singapore%20(for%20upload).pdf

National Research Foundation. (2008, June 24). *Opening Remarks By Dr Tony Tan, Chairman of the national research foundation, at the water leaders summit plenary forum, on "good governance, sustainable cities".* Retrieved from National Research Foundation: http://www.nrf.gov.sg/nrf/uploadedFiles/News_and_Events/SIWW%20(24%20June)%20(CAD).pdf

National Trade Union Congress. (2013). *RUN 350.* Retrieved 2013, from NTUC Income RUN 350: www.run350.com

Neo, H. (2007). Challenging the developmental state: Nature conservation in Singapore. *Asia Pacific Viewpoint*, 48(2), 186–199.

Ng, W., & Mendelsohn, R. (2005). The impact of sea level rise on Singapore. *Enviornment and Development Economics*, 10, 201–215.

O'Flaherty, J., & Liddy, M. (2018). The impact of development education and education for sustainable development interventions: A synthesis of the research. *Environmental Education Research*, 24(7), 1031–1049. https://doi.org/10.1080/13504622.2017.1392484

Ocal, A., Kisoglu, M., Alas, A., & Gurbuz, H. (2011). Turkish prospective teachers' understanding and misunderstanding on global warming. *International Research in Geographical and Environmental Education*, 20(3), 215. http://doi.org/10.1080/10382046.2011.588504

O'Connor, R. E., Bord, R. J., Yarnal, B., & Wiefek, N. (2002). Who wants to reduce greenhouse gas emissions? *Social Science Quarterly*, 83(1), 1–17.

Olsson, D., Gericke, N., & Chang-Rundgren, S.-N. 2016. "The effect of implementation of education for sustainable development in Swedish Compulsory Schools: Assessing pupils' sustainability consciousness." Environmental Education Research, 22(2), 176–202. http://doi.org/10.1080/13504622.2015.1005057

Ong, C. S. (2011) Yishun sec school students visit recycling station. Retrieved Aug 25, 2011, from http://www.tzuchi.org.sg/eng/index.php?option=com_content&view=article&id=1010:yishun-sec-school-students-visit-recycling-station&catid=116:2011news&Itemid=574

Padpadimitriou, V. (2004). Prospective primary teachers' understanding of climate change, greenhouse effect, and ozone layer depletion. *Journal of Science Education and Technology*, 13(2), 299–307.

Parry, M. L., Canziani, O. F., Palutikof, P. J., van der Linden, P. J., & Hanson, C. E. IPCC (2007): Summary for Policymakers. In M. L. Parry, O. F. Canziani, P. J. Palutikof, P. J. van der Linden, C. E. Hanson, M. L. Parry, O. F. Canziani, P. J. Palutikof, P. J. van der Linden, & C. E. Hanson (Eds.), *Climate Change 2007: Impacts, adaptation and vulnerability. Contribution of working group II to the fourth assessment report of the intergovernmental panel on climate change* (pp. 7–22). Cambridge: Cambridge University Press.

Pascua, L., & Chang, C.-H. (2015). Using intervention-oriented evaluation to diagnose and correct students' persistent climate change misconceptions: A Singapore case study. *Evaluation and Program Planning*, 52, 70–77.

Patton, M.Q. (2008). *Utilization-focused evaluation* (4th ed.).California: Sage.

Petersen, B., & Barnes, J. (2020). From hopelessness to transformation in geography classrooms. *Journal Of Geography*, 119(1), 3–11. https://doi.org/10.1080/00221341.2019.1566395

Popham, W. J. (1997). Consequential validity: Right concern—Wrong concept. *Educational Measurement: Issues and Practices*, 16, 9–13. https://doi.org/10.1111/j.1745-3992.1997.tb00586.x

Posner, G., Strike, K., Hewson, P., & Gertzog, W. (1982). Accommodation of a scientific conception: Toward a theory of conceptual change. *Science Education*, 66(2), 211–227. http://doi.org/10.1002/sce.3730660207

Powers, A. (2004). Teacher preparation for environmental education: Faculty perspectives on the infusion of environmental education into preservice methods courses. *Journal of Environmental Education*, 35(3), 3–11.

Pruneau, D., Khattabi, A., & Demers, M. (2010). Challenges and possibilities in climate change education. *US-China Education Review*, 7(9), 15–24.

Pruneau, D., Liboiron, L., Vrain , E., Gravel, H., Bourque, W., & Langis, J. (2001). People's ideas about climate change. A source of inspiration for the creation of educational programs. *Canadian Journal of Environmental Education*, 6, 121–138.

Punter, P., Ochando-Pardo, M., & Gracia, J. (2011). Spanish secondary school students' notions on the causes and consequences of climate change. *International Journal of Science Education*, 33(3), 447–464.

Ramesh, S. (2010, Mar 12). *National Climate Change Secretariat to be strengthened.* Retrieved Dec 28, 2010, from Channel News Asia: http://www.channelnewsasia.com/stories/singaporelocalnews/view/1043155/1/.html

Revelle, R., & Suess, H. (1957). Carbon dioxide exchange between atmosphere and ocean, and the question of an increase of atmospheric CO2 during the past decade. *Tellus*, 9(18), 18–27.

Reynolds, T. W., Bostrom, A., Read, D., & Morgan, M. (2010). Now what do people know about global climate change? Survey studies of educated laypeople. *Risk Analysis*, 30(10), 1520–1538.

Roberts, M. (2014). Powerful knowledge and geographical education. *The Curriculum Journal*, 25(2), 187–209. http://doi.org/10.1080/09585176.2014.894481

Rye, J., Rubba, P. & Wiesenmayer, R. (2007) An investigation of middle school students' alternative conceptions of global warming. *International Journal of Science Education*, 19(5), 527–551.

Salehi, S., Nejad, Z. P., Mahmoudi, H., & Burkart, S. (2016). Knowledge of global climate change: View of Iranian university students. *International Research in Geographical and Environmental Education*, 25(3), 226–243.

Saribas, D., Kucuk, Z., & Ertepinar, H. (2017). Implementation of an environmental education course to improve pre-service elementary teachers' environmental literacy and self-efficacy beliefs. *International Research In Geographical And Environmental Education*, 26(4), 311–326. https://doi.org/10.1080/10382046.2016.1262512

Sauvé, L. (2005). Environmental education: Possibilities and constraints. *Educação e Pesquisa*, 31(2), 317–322.

Schreiner, C., Henriksen, E., & Hansen, P. (2005). Climate education: Empowering today's youth to meet tomorrow's challenges. *Studies in Science Education* 41, 3–50.

Schultz, R., & DeMers, M. (2020). Transitioning from emergency remote learning to deep online learning experiences in geography education. *Journal of Geography*, 119(5), 142–146. https://doi.org/10.1080/00221341.2020.1813791

Sellan, R., Chong, K., & Tay, C. (2006). *Assessment Shifts in the Singapore Education System.* Retrieved Mar 30, 2008, from Singapore Examinations and Assessment Branch: http://www.iaea2006.seab.gov.sg

Sembawang Shipyard (2005) Sembawang Shipyard Private Limited 2005 Green Wave Environmental Care Competition. Retrieved June 1, 2013, from http://www.sembship.com/greenwave/

SENOKO Power (2009) National Weather Study Project. Retrieved June 1, 2013, from http://www.senokoenergy.com/nwsp/2009/NWSP%202009%20Inaugural%20Commemorative%20Magazine.pdf

Seow, T., & Ho, L. (2016). Singapore teachers' beliefs about the purpose of climate change education and student readiness to handle controversy. *International Research in Geographical and Environmental Education*, 25(4), 358–371. http://doi.org/10.1080/10382046.2016.1207993

SG Press Centre. (2008, Nov 5). *Transcript of Minister Mentor Lee Kuan Yew's dialogue at the Singapore energy conference.* Retrieved May 15, 2009, from SG Press Centre: http://www.news.gov.sg/public/sgpc/en/media_releases/agencies/mica/transcript/T-20081105-1.html

Shahid, Y., & Nabeshima, K. (2012). *Some small countries do it better: Rapid Growth and Its Causes in Singapore, Finland, and Ireland.* Washington DC: World Bank Publications.

Shepard, L. A. (1993). Chapter 9: Evaluating test validity. *Review of Research in Education*, 19(1), 405–450. https://doi.org/10.3102/0091732X019001405

Shepardson, D. P., Niyogi, D., Choi, S., & Charusombat, U. (2009). Seventh grade students' conceptions of global warming and climate change. *Environmental Education Research*, 15(5), 549–570. http://doi.org/10.1080/13504620903114592

Shulman, L. S. (1986). Those who understand: Knowledge growth in teaching. *Educational Researcher*, 67, 4–14.

Shumway, D., & Masser-Davidow, E. (1991). Disciplincarity: An introduction. *Poetics Today*, 12(2), 201–225.

Singapore Environment Council. (2022). School Green Awards. Retrieved June 2022 from https://web.sec.org.sg/sga/overview/

Singapore Examinations and Assessment Board. (2014). GCE O-Level Syllabus 2236. Retrieved 2016, from Singapore Examinations and Assessment Board: http://www.seab.gov.sg/content/syllabus/olevel/2016Syllabus/2236_2016.pdf

Singapore Green Plan. (2021). Retrieved Mar16, 2022, from https://www.greenplan.gov.sg/key-focus-areas/overview#green-economy

Singapore Institute of Biology (2008) Launch of climate change exhibition at the science centre. Retrieved from http://www.e2singapore.gov.sg/news_160508.html

Singapore School Examinations and Assessment Board. (2010). *GCE O-level syllabuses-combined humanities 2192.* Retrieved Jan 3, 2011, from Singapore School Examinations and Assessment Board.

Singapore School Examinations and Assessment Board. (2012a). *Geography GCE ordinary level (Syllabus 2236).* Retrieved Jan 16, 2013, from Singapore School Examinations and Assessment Board: http://www.seab.gov.sg/oLevel/2014Syllabus/2236_2014.pdf

Singapore School Examinations and Assessment Board. (2012b). *Chemistry: GCE ordinary level (Syllabus 5072).* Retrieved July 17, 2012, from Singapore School Examinations and Assessment Board: http://www.seab.gov.sg/oLevel/2013Syllabus/5072_2013.pdf

Singapore School Examinations and Assessment Board. (2012c). *Geography: GCE Ordinary Level (Syllabus 2235).* Retrieved July 17, 2012, from Singapore School Examinations and Assessment Board: http://www.seab.gov.sg/oLevel/2013Syllabus/2235_2013.pdf

Sinha, G., Smucker, T., Lovell, E., Velempini, K., Miller, S., Weiner, D., & Wangui, E. (2017). The pedagogical benefits of participatory GIS for geographic education. *Journal of Geography*, 116(4), 165–179. https://doi.org/10.1080/00221341.2016.1215488

Smith, J., di Sessa, A., & Roschelle, J. (1994). Misconceptions reconceived: A constructivist analysis of knowledge in transition. *Journal of the Learning Sciences*, 3(2), 115–163. http://doi.org/10.1207/s15327809jls0302_1

Smithwick, E., Baxter, E., Kim, K., Edel-Malizia, S., Rocco, S., & Blackstock, D. (2018). Interactive videos enhance learning about socio-ecological systems. *Journal of Geography*, 117(1), 40–49. https://doi.org/10.1080/00221341.2017.1374433

South West Community Development Council. (2007). *100,000 native plants at South West–Rooted in your community.* Retrieved Apr 16, 2008, from South West Community Development Council: http://www.southwestcdc.org.sg/1182064774425.html

Special Eurobarometer. (2009). *Europeans' attitudes towards climate change.* (W. 7.–T. Social, Ed.) Retrieved from European Commission: http://ec.europa.eu/public_opinion/archives/ebs/ebs_313_en.pdf

Steinberg, P. (1997). Political geography and the environment. *Journal of Geography*, 96(2), 113–118. https://doi.org/10.1080/00221349708978769

Stepans, J. (2003) *Targeting students' science misconceptions: Physical science concepts using the conceptual change model.* Tampa, FL: Showboard.

Stern, N. (2007). *The economics of climate change: The stern review.* Cambridge: Cambridge University Press.

Sternäng, L., & Lundholm, C. (2011). Climate change and morality: Students' perspectives on the individual and society. *International Journal of Science Education*, 33(8), 1131–1148.

Stigler, J., & Hiebert, J. (2009). *The teaching gap: Best ideas from the world's teachers for improving education in the classroom* (pp. 1–224). New York: Simon and Schuster.

Stobart, G. (2001). The validity of national curriculum assessment. *British Journal of Educational Studies*, 49(1), 26–39.

Stoltman, J. P. (1997) The international charter on geographical education: Setting the curriculum standard. *Journal of Geography*, 96(1), 3. http://doi.org/10.1080/00221349708978752

Summers, M., Childs, A., & Corney, G. (2005). Education for sustainable development in initial teacher training: Issues for interdisciplinary collaboration. *Environmental Education Research*, 11(5), 623–647.

Sund, P., & Gericke, N. (2020). Teaching contributions from secondary school subject areas to education for sustainable development – A comparative study of science, social science and language teachers. *Environmental Education Research*, 26(6), 772–794. https://doi.org/10.1080/13504622.2020.1754341

Sung, J. (2015). Climate change education and education for sustainable development in the Republic of Korea: A status report. *Journal of Education for Sustainable Development*, 9(1), 78–89. http://doi.org/10.1177/0973408215569116

Svihla, V., & Linn, M. C. (2012). A design-based approach to fostering understanding of global climate change. *International Journal of Science Education*, 34(5), 651–676.

Taber, F., & Taylor, N. (2009). Climate of concern – A search for effective strategies for teaching children about global warming. *International Journal of Environment & Science Education*, 4(2), 97–116.

Tan, G. I., Lee, K. C., & Goh, K. (1998). Survey of environmental knowledge, attitudes and behaviour of students in Singapore. *International Research in Geographical & Environmental Education*, 7(3), 181–202.

Tan, I., & Chang, C. (2008). Geography education for sustainable development in Southeast Asia. *International Research in Geographical and Environmental Education*, 17(4), 289–291.

Tan, K. C. D., Goh, N. K., Chia, L. S., Treagust, D. F. (2002). Development and Application of a Two-tier multiple choice diagnostic instrument to assess high school students' understanding of inorganic chemistry qualitative analysis. *Journal of Research in Science Teaching*, 39(4), 283–301.

Tan, S., Wirtz, J., Jung, K., & Kau, A. (2001). attitudes toward work, pecuniary adherence, materialism, feminism, environmental consciousness, and media credibility. *Singapore Management Review*, 23(1), 59–86.

Teh, J. (2008, Sep 17). Sea levels may rise: But Singapore's okay. *The New Paper*.

Teng, A. (2020, Oct 20). Singapore students familiar with global issues, But know Less about world conflicts and economy: 2018 Pisa study. Retrieved Aug 24, 2021, from https://www.straitstimes.com/singapore/parenting-education/local-students-know-more-about-climate-change-than-world-conflicts

Teo, P. (2004). "Clean and green — That's the way we like it": Greening a country, building a nation. *Journal of Language and Politics*, 3(3), 485–505.

The Belgrade Charter. (1975, Oct 22). *The Belgrade charter: A framework for environmental education*. Retrieved June 21, 2011, from UNESCO: http://unesdoc.unesco.org/images/0001/000177/017772eb.pdf

The Heartland Institute. (2007). *Climate change in the classroom education or indoctrination? Part 2: Scientific critiques of global warming alarmism.* Retrieved from The Heartland Institute: http://heartland.org/sites/all/modules/custom/heartland_migration/files/pdfs/24283.pdf

The Straits Times. (2007, Mar 7). 2-year study to gauge impact of global warming. *The Straits Times*, p. 39.

The Straits Times. (2008, Jan 22). Climate change deal: Jaya sets out S'pore's guiding principles. *The Straits Times*, p. 30.

The Straits Times. (2010, June 16). *Flash floods wreak havoc.* Retrieved Feb 2, 2011, from The Straits Times: http://www.straitstimes.com/BreakingNews/Singapore/Story/STIStory_541057.html

Tippett, C.D. (2010). Refutation Text in Science Education: A Review of Two Decades of Research. *International Journal of Science and Mathematics Education*, 8, 951–970. https://doi.org/10.1007/s10763-010-9203-x

Today. (2012, Mar 14). *Climate change authority responds to WWF's 'largest carbon footprint' charge.* Retrieved Mar 16, 2012, from Today: http://www.todayonline.com/Singapore/EDC120314-0000091/Climate-change-authority-responds-to-WWFs-largest-carbon-footprint-charge

Tomlins, B., & Froud, K. (1994) Environmental education: Teaching approaches and students' attitudes: A briefing paper. Slough: National Foundation for Educational Research.

Trajber, R., & Mochizuki, Y. (2015). Climate change education for sustainability in Brazil: A status report. *Journal of Education For Sustainable Development*, 9(1), 44–61. http://doi.org/10.1177/0973408215569113

Treagust, D. F. (1988). Development and use of diagnostic tests to evaluate students' misconceptions in science. *International Journal of Science Education*, 10(2), 159–169.

Treagust, D. F., & Duit, R. (2008). Conceptual change: A discussion of theoretical, methodological and practical challenges for science education. *Cultural Studies in Science Education*, 3, 297–328.

Trumper, R. (2010). How do learners in developed and developing countries relate to environmental issues? *Science Education International*, 21(4), 217–240.

Tsui, C. Y., & Treagust, D. F. (2010). Evaluating secondary students' scientific reasoning in genetics using a two-tier diagnostic instrument. *International Journal of Science Education*, 32(8), 1073–1098.

UNESCO. (2010). *UNESCO strategy for the second half of the United Nations decade of education for sustainable development.* Paris: UNESCO.

UNESCO. (2012). *Education for sustainable development.* Retrieved May 25, 2013, from United Nations Organization for Education, Science and Culture: http://www.unesco.org/new/en/education/themes/leading-the-international-agenda/education-for-sustainable-development/education-for-sustainable-development/

UNESCO (2021). Climate Change Education, Retrieved 15 June 2022 from UNESCO: https://www.unesco.org/en/education/sustainable-development/climate-change

UNESCO & UNEP. (2011). Climate change starter's guidebook: An issues guide for educational planners and practitioners.

Ungar, S. (1992). The rise and (relative) decline of global warming as a social problem. *The Sociological Quarterly*, 33(4), 483–501.

United Nations Framework Convention on Climate Change. (2013). *Kyoto Protocol.* Retrieved June 1, 2013, from United Nations Framework Convention on Climate Change: http://unfccc.int/kyoto_protocol/items/2830.php

van den Broek, P., & Kendeou, P. (2008). Cognitive processes in comprehension of science texts: The role of co-activation in confronting misconceptions. *Applied Cognitive Psychology*, 22(3), 335–351. https://doi.org/10.1002/acp.1418

Van Doren, M. (1959). *Liberal education.* MA: Beacon Press.

Van Petegem, P., Blieck, A., Imbrecht, I., & Van Hout, T. (2005). Implementing environmental education in pre-service teacher training. *Environmental Education Research*, 11(2), 161–171.

Voltz, D. L., Sims, M. J., & Nelson, B. P. (2010). *Connecting teachers, students, and standards: Strategies for success in diverse and inclusive classrooms.* Alexandria, VA: ASCD.

Von Aufschnaiter, C., & Rogge, C. (2010). Misconceptions or missing conceptions. *Eurasia Journal of Mathematics, Science & Technology Education*, *6*(1), 3–18.

Vosniadou, S. (2003). Exploring the relationships between conceptual change and intentional learning. In G. M. Sinatra & P. R. Pintrich (Eds.), *Intentional conceptual change* (pp. 377–406). Mahwah, NJ: Lawrence Erlbaum Associates.

Vosniadou, S. (2007). The conceptual change approach and its re-framing. In S. Vosniadou, A. Baltas, & X. Vamvakoussi (Eds.), *Reframing the conceptual change approach in learning and instruction* (pp. 1–15). Elsevier Science.

Vosniadou, S. & Verschaffel, L. (2004). Extending the conceptual change approach to mathematics learning and teaching. *Learning and Instruction*, 14, 445–451.

Vygotsky, L. (1986). *Thought and language* (A. Kozulin, Ed., & A. Kozulin, Trans.) Cambridge: MA: The MIT Press.

Wade, K. (1996). EE teacher inservice education: The need for new perspectives. *Journal of Environmental Education*, 27(2), 11–17.

Wang, J. R. (2004). Development and validation of a two-tier instrument to examine understanding of internal transport in plants and the human circulatory system. *International Journal of Science and Mathematics Education*, 2(2), 131–157.

Wee, B. (2001). *Urban farmers: Growing grassroots.* Singapore: The Straits Times.

Wee, B. (2008). Moving toward sustainability? The face of environmental education in Singapore. *Green Teacher* 83, 35–38.

Wee, B., Harbor, J., & Shepardson, D. (2006). Multiculturalism in environmental science: A snapshot of Singapore. *Multicultural Perspectives*, 8(2), 10–17.

Wi, A. (2019) Citizen participation as a key enabler for successful public education policies in climate change mitigation in Singapore. *International Research in Geographical and Environmental Education*, 28(1), 53–69. http://doi.org/10.1080/10382046.2018.1430789

Wi, A., & Chang, C.H. (2019) Promoting pro-environmental behaviour in a community in Singapore – from raising awareness to behavioural change, *Environmental Education Research*, 25(7), 1019–1037. http://doi.org/10.1080/13504622.2018.1528496

Wiggins, G., & McTighe, J. (2005). *Understanding by design.* Alexandria, VA: Association for Supervision and Curriculum Development.

Wikipedia. (2013, July 17). *Education.* Retrieved July 20, 2013, from Wikipedia the free encyclopedia: http://en.wikipedia.org/wiki/Education

Williams, P. (2008). Leading schools in the digital age: A clash of cultures. *School Leadership & Management*, 28(3), 213–228.

Wilson, S., & Wineburg, S. (1988). Peering at history from different lenses: The role of disciplinary perspectives in the teaching of American history. *Teachers College Record*, 89, 525–539.

Wise, S. B. (2010). Climate change in the classroom: Patterns, motivations, and barriers to instruction among Colorado science teachers. *Journal of Geoscience Education*, 58(5), 297–309.

Wong, B. F., & Stimpson, P. (2003). Environmental education in Singapore: A curriculum for the environment or in the national interest? *International Research in Geographical and Environmental Education*, 12(2), 123–138.

Wong, P. (1992). Impact of sea-level rise on the coasts of Singapore: Preliminary observations. *Journal of Southeast Asian Earth Sciences*, 7(1), 65–70.

Wong, W. (2012, July). *Yearbook of Statistics Singapore, 2012.* Retrieved May 15, 2012, from Department of Statistics: http://www.singstat.gov.sg/publications/publications_and_papers/reference/yearbook_2012/yos2012.pdf

Woo, S., & Lee, A. (2013, June 25). *Hail hits Singapore.* Retrieved June 25, 2013, from Today Online: http://www.todayonline.com/singapore/hail-hits-singapore

Yang, X., Wei, L., & Su, Q. (2020). How is climate change knowledge distributed among the population in Singapore? A demographic analysis of actual knowledge and illusory knowledge. *Sustainability*, 12(9), 3782. http://doi.org/10.3390/su12093782

Yazdanparast, T., Salehpour, S., Masjedi, M. R., Seyedmehdi, S. M., Boyes, E., Stanisstreet, M., & Attarchi, M. (2013). Global warming: Knowledge and views of Iranian students. *Acta Medica Iranica*, 51(3), 178–184.

Yee, J. (2020). Environmental education in Singapore: An analysis of environmental knowledge in the lower secondary geography curriculum. *HSSE Online*, 9(1), 24–38.

Young, M. (2008) From constructivism to realism in the sociology of the curriculum. In G. J. Kelly, A. Luke, & J. Green (Eds.), *Review of research in education* (Vol. 32, pp. 1–28). Thousand Oaks, CA: Sage.

Index

Pages in *italics* refer figures and pages in **bold** refer tables.

For Product Safety Concerns and Information please contact our EU
representative GPSR@taylorandfrancis.com
Taylor & Francis Verlag GmbH, Kaufingerstraße 24, 80331 München, Germany

www.ingramcontent.com/pod-product-compliance
Lightning Source LLC
LaVergne TN
LVHW021124300926
849160LV00010B/858